Pain

MAPS OF THE MIND

Steven Rose, General Editor

MAPS OF THE MIND

Steven Rose, General Editor

—

Pain

THE SCIENCE OF SUFFERING

Patrick Wall

Columbia University Press

New York

Columbia University Press
Publishers Since 1893
New York   Chichester, West Sussex

Library of Congress Cataloging-in-Publication Data
Wall, Patrick D. (Patrick David), 1925–
Pain : the science of suffering / Patrick Wall.
    p. cm. — (Maps of the mind)
Includes index.
ISBN 978-0-231-12006-7 (cloth : alk. paper)—ISBN 978-0-231-12007-4 (pbk. : alk. paper)
1. Pain.   I. Title.   II. Series.
RB127.W355 2000
616′.0472—dc21

                                                    00–023889

☉

Columbia University Press books are printed on permanent and durable acid-free paper.
Printed in the United States of America

Contents

Preface

This book will explore all we know now about the nature and causes of pain. There is good reason to do this because we have all experienced pain and are puzzled by it. We all fear for the future. Furthermore, we have all witnessed others in pain and have wondered in sympathy at their experience and our often frustrated attempts to help them. We will begin with the surprising facts about what people and animals do and experience when their bodies are damaged. We need to start with this because we have to clear our minds of the idea of a rigid, simple signaling system. It is simply not true that a particular injury generates a fixed amount of pain, and that feeling pain means injury. We shall then divert briefly to the confused mess generated by philosophers who constructed idealized "rational" sensory systems in happy ignorance of the working of living organisms. We will rapidly return to the real world of our nervous systems and how we and others react to our injuries, pain, and suffering. The rationale of successful therapies will be reviewed, as will the nature of intractable pains, which still defeat everyone. Finally, we shall try to fit these facts into a scheme by which the body and brain bring together all the relevant data to make an overall posture that generates our sense of ease or distress.

My desire to write this book comes from my own history. More than fifty years ago, as a medical student, I began to see patients in pain and realized that the explanations given to them and to me by my teachers were overt rubbish. The fantasy explanations often depended on mechanical disorders for which there was no evidence, such as trapped nerves, extra ribs, strained muscles, or floating kidneys. If those failed to convince even the doctors, there was a leap to using as an explanation the supposed inadequate personalities of the patients: neurosis, hypochondria, hysteria, and malingering.

Over forty years ago, I began, as a neuroscientist, to examine the conducting nervous system of animals. It was immediately apparent that the predicted hard-wired, line-dedicated, specialized pain system did not exist. Rather, there is a subtle multiplexed reactive system that informs us simultaneously about events in the tissues and in the thinking parts of the brain. As a consequence of these studies, it became apparent that the separation of sensation from perception was quite artificial and that sensory and cognitive mechanisms operated as a whole. Since that time, a large number of highly inventive scientists, psychologists, and clinicians have joined to create a contemporary picture of pain mechanisms. Last, like anyone, I have a personal concern with my own pain, in my case dominated by a widespread cancer (which is for the time being responding to therapy).

This is not a textbook. There are many such books, including *The Challenge of Pain* (Penguin, London, 1999), written by myself and Ron Melzack, and the fourth edition of the *Textbook of Pain* (Churchill-Livingstone, Edinburgh, 1999), which Melzack and I edited. Neither is this book a do-it-yourself recipe book for pain relief. It is intended to give the readers the power of understanding processes in their own body.

I am indebted to the advice of friends: Peter and Nehama Hillman, Sue Draney, Julia and Robert Kibblewhite, Steven Rose, the series editor, Peter Tallack, the editor, and, of course, Mary Helton.

This book reflects the awakening of study and concern about pain that has been growing for twenty years. I am in awe of the group of younger scientists who have taken up the challenge. I hugely admire the new breed of clinicians who are leading treatment out of the barren desert of ignorance and neglect. Above all, this book is inspired by those who suffer and those who care. They are our ultimate teachers. This book is created both by and for them.

Pain

Private Pain and Public Display

It is crucial that we begin with precise and objective reports of what people and animals do when injured. The reports do not match the expectation of the victim or of the observer. In exploring the nature of pain, it will be necessary to separate reality from what we think ought to be observed. We will start with sudden events where a previously "normal" being is abruptly converted to a "sick" one. Of course, no such event occurs in a vacuum, as there is always a surrounding scene and the victim arrives at the accident with a personal and genetic history. Later, we will have to incorporate the vastly more common and less dramatic situation in which the onset of disease and pain is insidious.

A Swiss Army Officer

A forty-three-year-old reserve major, described by his wife as tough and taciturn, was skiing with his squad in the Upper Engadine region of Switzerland when a snow bridge collapsed below him. He remembers free falling into a crevasse with ice walls in front and behind, and scraping down one wall. With a tremendous crash and thump, he found himself wedged firmly in the ice crack. One arm was jammed above his head and he could not move his legs. He remembers hearing his gun and ski poles rattling down below, deeper into the crevasse. He was winded but was surprised to feel no pain whatsoever. He looked up and saw his men peering over the brink, and called out that he was all right but could not move.

A man was lowered down to him on a line and put a sling around him. The men above hauled on the ropes, and he remembers his relief on feeling himself swaying free. They carried him down to an open area and radioed for a helicopter, which arrived after twenty-five minutes. His men were unusually quiet and subdued, in contrast to their normal boisterous behavior. He recalls feeling ashamed, as he had lectured on how to avoid such accidents. He wondered what this would do to his chances of promotion, and discussed this with the sergeant, who tried to cheer him up. During all this time, he recalls no trace of pain, either when he hit the ice slot or during his rescue.

He was strapped onto a stretcher and the helicopter took off. Just then, some forty-five minutes after his fall, a searing pain started in his left shoulder and spread to his neck and chest. He cried out. A crew member gave him a subcutaneous injection of fifteen milligrams of the narcotic morphine from the standard equipment in the emergency kit. By the time they arrived in hospital, he was dozy and the pain had lessened.

In the hospital, it was found that he had a dislocated left shoulder, a broken left collar bone, and serious bruises over his pelvis and upper legs. He was briefly anesthetized, and the shoulder bones were put in their proper place. He was put to bed and slept. Next morning, his shoulder ached all the time and he felt severe stabs of pain if he moved. He was sore all over, and was given painkillers. He felt exhausted and dozed for long periods. When the doctors came on their routine ward round, his pain escalated as they uncovered him and he cried out when they gently touched his shoulder. For the rest of the day, he curled up, moving as little as possible. He wanted no food. When visitors came, he put on his standard act: "Nothing to it," "Just a bit of a fall," and "I'll be out of here in a day or two." Within himself, he wished they would go away and leave him alone.

This history has two clear epochs. In the first emergency period of forty-five minutes, where survival, escape, and rescue had clear priority, there was injury but no pain. He was mentally clear and supervising his own rescue. Furthermore, he was assessing the situation clearly but blaming himself and fearing for the future. In the second period, when pain began, recovery from the injury had priority. Pain was present and increased with movement or touch. Beyond the presence of pain, his usual character had changed: normally a very active man, he was overwhelmed by lethargy and fatigue; usually a good eater, he had no appetite; habitually gregarious, he disliked company, although verbally he put on a very good act in imitating his old self. Within himself, he displayed the complete syndrome of the best tactics for recovery in people or animals: don't move, and don't let anyone else move you, just sleep. Outside himself, he displayed the opposite, for the benefit of other people and for his own image: "I'm all right," "Soon be out," "Don't worry," "It only hurts when I laugh."

The Anzio Beachhead

During the winter of 1943–44, Allied troops came to a halt in their advance up Italy on the Gustav line, which included the slaughter point of Monte Cassino. In an attempt to outflank this line, American and British troops landed 50 miles north on the beaches of Anzio in January 1944. They landed successfully on the coastal strip but were trapped when the Germans regrouped in the hills. It took until May 1944 before there was a breakout and Rome was captured. During this time, the Allied troops' lines hardly moved and they suffered heavy casualties, mainly from persistent artillery fire.

Harry K. Beecher was the medical officer admitting casualties to one of the few hospitals on the beachhead. He was later to become a leader in the new clinical research on pain as professor of anesthesia at Harvard. His concern for humanity reflected that of his ancestor, Harriet Beecher Stowe, author of *Uncle Tom's Cabin*. Every wounded man who could speak was asked the same question: "Are you in pain? Do you want something for it?" All of these men were seriously wounded, as they had already been sent back after first aid in advanced medical posts. Some of these men were in an exalted state, recognizing their near brush with death. Beecher collected the answers and was astonished that 70 percent of the men answered no to both questions. When the war was over, Beecher asked the same questions of an age-

FIGURE 1. Traditional drawing of the battle between the Imam Ali, cousin of Mohamed, and Amr Bin Wid, champion of the enemy host. Amr Bin Wid threw his amputated leg at Imam Ali before being killed. The two-pointed sword called Al-Sahar remains on display in the Topkapi Palace in Istanbul.

matched group of civilian men who had been operated on at the Massachusetts General Hospital in Boston; 70 percent answered yes to both questions.

Beecher reasonably concluded that something about the situation in which the tissue damage was inflicted influenced the amount of pain suffered. Beecher had a theory about what was the crucial difference in the two situations. To be wounded on the Anzio beachhead had a positive biological advantage over not being wounded. This paradoxical statement needs explanation. To be wounded and to reach the hospital at Anzio implied a good chance of evacuation and survival. To remain in the line unwounded implied a serious risk of being killed. Of the 767 men in the American Rangers battalion who attacked in the attempted breakout on January 30, only six returned. Beecher proposed that the men he questioned in Anzio were pain free because they were in an exalted state in the expectation of survival with honor. Therefore, he suggested that there are rare circumstances in which wounding is advantageous, and those wounds are pain free. I doubt this reasoning, but there is no doubt that he observed numbers of seriously wounded men who were not in pain.

The Yom Kippur War

In October 1973, Syria and Egypt attacked Israel and there was a brief, violent war. Dr. Carlen from Canada, Dr. Bill Noordenbos from Holland, and I decided to study a complete sample of Israeli soldiers who had suffered a traumatic amputation during that war. We examined seventy-three amputees from days to months after their injury. Their mean age was twenty-six and ranged from nineteen to forty-five years. This study of the subsequent history of amputees in or out of pain was possible because all Israeli amputees attended a single rehabilitation hospital at Tel Hashomer outside Tel Aviv. On questioning about their first sensation at the time of injury, the great majority clearly described their sensations with neutral words such as "bangs," "blows," and "thumps." Not one described a flash of pain that then died down. These men uniformly expressed their surprise at not feeling pain, often beginning with "Doctor, you won't believe it but . . ." A minority had felt pain from the first moment.

With Beecher's reasoning in mind, we began to inquire about the precise circumstances of their wounding. The scene at Anzio was evidently one of unremitting horror and terror, continuing night and day, with the common soldier adjusting to some passive tactic that he hoped would end in survival. For those of us fortunate enough not to have been in battle, we perhaps imagine that painless injuries could occur only in the heat of combat when the "blood was up" and the victim was engaged in some intense action. But, far from this picture of continuous intensity, the Yom Kippur War was often intermittent, scattered with abrupt short violence. Some of the men had suffered in road accidents as they raced at night with lights out on unfamiliar roads. Some had been asleep in a quiet area when they were hit by an unexpected long-range shell. Others had been hit suddenly by the accidental firing of weapons from their own side. It was evident that painless injury could occur in men who were in no unusual state of mind. The episode appeared to begin precisely upon impact and did not depend on some prior expectation.

Our team, which included fluent Hebrew speakers, very tentatively and diplomatically explored the question whether any of these men greeted their injury as welcome. There was never a hint that anyone adopted the Darwinian approach that being wounded increased their chance of survival. No soldier reported even a fleeting sense of relief that he had escaped alive from the killing fields. Elsewhere, I spoke to a man from another army who had shot off one of his own toes in order to get out of action. He said it hurt badly and immediately. Yet the overwhelming reaction of these seventy-three soldiers to their wound was anger. Surprisingly, it was often directed at themselves: "If only I had not gone into that house" or "If only I had not climbed out of the trench."

I found the most bizarre story of this kind to come from a man who had lost three fingers from one hand. He had been standing head and shoulders out of a tank turret when he saw an Egyptian wire-guided antitank missile streaking towards him. He dodged down, leaving his hands on the rim of the turret so that he lost his fingers when the missile exploded. He said, "What a fool I was. If I had time to get my head out of the way, I certainly had time to move my hands." Next to themselves, they blamed officers and, only low on the list, the enemy, who had, after all, really been responsible.

We can leave the topic of emergency painless injury as a fact that we must accept and explain without reverting to ad hoc attempts to classify it as a very special case, as Beecher did, or by using meaningless terms such as *shock*—the victims had clear minds and were behaving rationally. We have all witnessed one such episode on television. President Ronald Reagan was shot with a 9-mm bullet that entered his chest as he walked from a Washington hotel. He was slammed roughly into his car by the Secret Service men. He and the others in the car did not know he was wounded. He began to feel unwell, and there was a discussion about the possible damage to him when he slumped against the car door. On the anniversary of the shooting, Reagan appeared on a

FIGURE 2. President Ronald Reagan seconds after being shot in the chest with a 9-mm bullet on March 30, 1981, outside a Washington hotel. He felt no pain and did not know he had been shot until he began to feel faint as blood poured into his chest cavity while in the car racing from the scene.

CBS documentary and said, with his wondrous command of English: "I had never been shot before except in the movies. Then you always act as though it hurts. Now I know that does not always happen."

So much for the men's reports of their immediate sensations. What was their sensory state after some weeks when we saw them? Within twenty-four hours of their amputation, 65 percent experienced a "phantom limb," a name given by the American Civil War physician Wier-Mitchell to the clear sensation that the missing limb is still present. The remaining 35 percent all felt a phantom limb within a few weeks. Pain in the phantom was experienced by 67 percent, who described it using the words "jabs," "strong current," "pins and needles," "burning," "knifelike," "pressure," "cramps," "crushing," and "vicelike." In addition, many had pain in their stumps, with or without phantom pain as well. On careful examination of the stumps, it was found that every man had at least one area of intensely painful hypersensitivity. At the time these men were examined, in 1973 and 1974, 80 percent had stumps that appeared perfectly healed with no signs of infection. It is particularly sad to report that when this same group of men was examined fifteen years later, even though all signs of infection had gone and all stumps appeared perfectly healed, the pain reports were identical to those we had reported soon after the war.

Now we have three new problems to absorb and explain. What can be the explanation of the common report of no pain at the time of the injury but pain within a day? Second, some of the pains appeared to the victims to be located in a lost limb. And third, some of the pains persisted even when there seemed to be complete healing of the remaining damaged tissue.

Animals with Abrupt Injuries

The major horse race in Britain at the end of the season is the Epsom Derby. To everyone's surprise, it was won in 1980 by a horse called Henbit, who was far from being the favorite. Half a mile from the end of the race, Henbit was running in the middle of the pack when it stumbled into a hole. The jockey felt and heard a crack. The horse accelerated away from the pack and, with the spectacular perfect gait of a thoroughbred at full gallop, won the race. In the paddock, the jockey leapt off and went to feel the right foreleg. As he suspected, and as was later confirmed by X-ray, the cannon bone, a fine long bone, had been fractured when the horse stumbled. The next day's newspapers had a headline that read "Gallant Henbit's career may end." The horse said nothing but began to limp. With good veterinary care, the fractured cannon bone appeared to knit perfectly, but something changed in the horse. They never got it up to speed again and it retired to stud. Smart horse?

Henbit is not the only horse to have impressed people with its "courage." In 1632, Sweden's King Gustav Adolf II was killed at the battle of Lützen in Poland. His horse had a large open gunshot injury in its left shoulder. In spite of this, the horse carried his dead master off the battlefield and walked to the Baltic coast where it died. Awed by this display of "valor," the Swedes transported king and horse back to Stockholm. The horse, now stuffed, stands today in the stables below the royal palace.

Deer hunters report the effect of rifle shooting at one member of a herd of deer. At the sound of the shot, the entire herd takes off at full

FIGURE 3. Henbit races away from the pack of horses to win the 1980 Epsom Derby. Henbit had stumbled 150 yards before this picture and had broken the cannon bone in its right foreleg. The picture shows the horse taking its full weight on the leg containing the broken bone.

speed, aiming for cover. One of the herd has been hit but, with anything short of an immediately lethal injury, it is impossible to identify the wounded animal from the others by its speed, gait, or skill. Long after, the wounded animal may be seen again, curled up, separated from the herd and drowsy.

Dog owners have often seen their gentle domesticated friends suddenly display their hoodlum characters given the chance of a dogfight. Fur flies, canine teeth puncture skin, and chunks of flesh are ripped free. Does the wounded dog stop and surrender? You wish it would. When finally separated from the brawl and taken home, the wounded dog again changes its character. It lies curled up, quiet, sleepy, wants no food, and licks its wounds. Be very cautious in examining the wounds because the dog is likely to yelp and bite you.

The subject of hunting is now under intense scrutiny for many good reasons. A recent study of stags chased and killed by hounds showed that their blood chemistry was grossly abnormal when compared with animals abruptly and humanely slaughtered. The changes were characteristic of intense stress, fatigue, and excessive exercise. Similar but less marked changes are seen in marathon runners who have voluntarily put themselves through a grueling race. The ethical question of cruelty rests on whether anyone is justified in putting an animal through this involuntary stress and not on the unanswerable question of what the animal felt at the moment of death. The fact that a dying soldier on a battlefield may not feel pain in his last moments does not remove the stress from which he suffers, nor does it solve the ethical question of whether his shooting was permissible.

I stress here the similarities between stressed and wounded humans and animals. That does not mean identity between species. There are profound differences. For example, when an old deer is culled by shooting and drops dead, the other members of the herd briefly startle but then continue grazing and ignore the corpse. Deer are evidently not human, but that does not give open permission to kill deer.

Animals and humans in the early stages after abrupt injury may appear to ignore the injury. They proceed with an activity that has a higher priority than care for the wound. Unfortunately, this common observation has been interpreted by some hunters and others in charge of animals to mean that animals do not feel pain. Examination of the victim a few hours later quickly dispels that generalization because, by then, they show all the same signs as humans who are trying to recover from injury.

Furthermore, there are long-term consequences wherever they have been examined. Animals learn and become cautious and skilled in avoidance. Henbit would never gallop at full speed despite the apparently perfect healing of its fracture. When the wily, fat old pike, who drives generations of anglers to distraction because he will not take the bait, finally gives up to senility, it is found that his mouth has the scar of an old hook from which he escaped and learned.

Hospital Emergency Rooms

My friend Ron Melzack, a Canadian psychologist, and I decided to examine and compare patients who were admitted to the emergency room (the casualty department) of the largest general hospital in Montreal. In a gentler setting, we could probe with questions that had not been possible in more urgent and dramatic scenes. We examined the first 138 patients to enter, who were alert, rational, and coherent, 37 percent of whom said they did not feel pain at the time of the injury. Of those patients with injuries limited to their skin, such as abrasions, cuts, and burns, 53 percent had a pain-free period. However, of those patients with deep tissue injuries, such as fractures, sprains, and stabs, only 28 percent had a pain-free period. The majority of these people reported the onset of pain within an hour, although some did not feel pain for many hours. The predominant emotions of the patients were embarrassment at appearing careless or worry about loss of wages. None expressed any pleasure or indicated any prospect of gain as a result of the injury.

A fifty-two-year-old senior machine shop foreman lay on a gurney in the emergency room after a collapse of heavy machinery had amputed the front of his right foot. He stated that there was no pain. This was not his first experience of painless sudden injury because an unexploded aircraft canon shell had lodged in the upper part of his leg during the Second World War, and we observed the old scar. He was coherent, sad, and thoughtful, and said, "What a fool they will think I am to let this happen" and "There goes my holiday." He lay still on the trolley with an intravenous drip running while waiting to go to the operating room. After a while he complained of a painful cramp in his left leg although the injured leg remained pain free. The pain went away with massage. Evidently, his analgesia was present only in the region of the original injury. This phenomenon had already been reported by Beecher, who observed that pain-free casualties complained of pain when intravenous needles were inserted.

We can now summarize key points of sudden injury. For a start, sudden injury may or may not be painful. The victims can be coherent and rational throughout. There may be no pain from the moment of injury. The pain-free state is localized precisely to the site of the injury. And all victims are eventually in pain.

Let us now turn to the majority (63 percent) who were in pain from the moment of injury. How much pain? They were asked to rate their pain on a scale of 0 to 10, with 10 as the worst pain imaginable. The answers were widely scattered. Anyone, expert or not, observing someone who is injured almost inevitably assigns an "appropriate" amount of pain that they expect. On what do they base this assignment of the appropriate: personal experience, professional experience, empathy, sympathy, knowledge of the victim? The staff of the emergency room thought that 40 percent were making "a terrible fuss," nearly 40 percent were "denying" pain, and 20 percent gave the "appropriate" answer. It is obvious there is something fundamentally wrong here. People generally are convinced that a certain degree of injury inevitably produces and justifies an appropriate amount of pain. Clearly this is not the case, but we have great difficulty in accepting the fact. (It is strange that even professionals may ignore their experience and persist in expecting patients to display only an appropriate amount of pain.)

A major theme of this book will be the exploration of the factors that, in addition to overt damage, produce pain and modulate its intensity. One crucial aspect is that patients are not only assessing their private misery but also making a public display. Their private misery is not necessarily about the pain. For example a twenty-two-year-old Israeli Army woman lieutenant with one leg blown off above the knee by a shell explosion was in deep distress with tears flooding over her face. When asked about her pain, she replied, "The pain is nothing, but who is going to marry me now?"

What did the patients say about their pain? Melzack has made an extensive study of the words people use, as will be discussed in chapter 2. He divides the words into "sensory," such as *sharp, burning,* and *stinging,* which describe the sensation itself, and into "affective," such as *tiring, sickening,* and *annoying,* which describe what the feeling is doing to the person. It is interesting that on their first encounter in the emergency room with doctors and nurses, patients used, almost entirely, the sensory words. Much later they would add the affective words. In this emergency situation, the first priority of communication was to inform those who brought aid exactly those details they would need to diagnose the injury. They delayed, to less urgent times, the information about what the injury was doing to their mood.

Torture

Governments have refined techniques over the centuries for deliberately inflicting pain. The victim is in a unique situation quite unlike that of any injury or disease because he is helpless and there will be no help. If the torture is part of an interrogation, the only way to end the pain is to tell the interrogator what he wants to hear or, more likely, to invent false yet plausible information. A few victims may have the skill to dissociate themselves from the present and to enter a fantasized world. The vast majority react in the predicted way and search only for ways to end the torture. Where the torturer is a sadist or has the intention to terrorize the population at large, the victim loses even the option that confession will stop the torture.

Christopher Buney wrote of his own interrogation in a German prison:

> Suddenly the major turned and strode across the room and struck me in the face with a swing of his open hand. I knew now what was to come. The first impact still took me by surprise. There is a sense of shame following an unanswerable blow which has nothing of fear in it but which is more demoralizing than any pain.

To experience helplessness in the face of impending death wipes out confidence. The horror of this state becomes primary beyond the misery of the pain. A South African doctor conscripted into the army describes being presented with Namibian prisoners under torture and being ordered to treat them so that interrogation could continue. He gave morphine to ease their pain. I believe this compounds the atrocity. I do not believe that it was ethically permissible for the doctor to relieve pain where the consequence was that violent beating would continue to the point of threatening the prisoner's life.

Individuality can be demolished without pain. In the early 1970s, the British Army in Northern Ireland introduced a new high-tech method of interrogation without pain. Arrested men were made to lean at 45 degrees, supported by their handcuffed hands on a wall. A bag was placed over their heads so they could see nothing. Intense noise from loudspeakers prevented hearing. If they collapsed, they were propped up again. At irregular intervals, they were taken out and interrogated but were otherwise left in their unmoving posture of sensory isolation for days. When these men were examined, long after their release, many remained broken zombies, apathetic, tremulous, and unable to function. Primo Levi wrote of his experience in Auschwitz: "Anyone who has been tortured, remains tortured." The British government set up a judicial committee, the Gardner Commission, to investigate torture, and the practice was forbidden. For many of us, the fear of our manner of dying is much greater than our fear of death itself.

Masochists

It would be narrow-minded to avoid admitting that there is a small fraction of any community that invites the very pain that the great majority attempt to avoid or cure. One such group are the athletes and aerobic buffs whose motto is "No pain, no gain." We can understand if not applaud their conviction that pain is a measure of achievement and is therefore welcome. A quite different group are those in the quasi-legal underground sadomasochist subculture who seek pleasure in pain.

An interview with an attractive woman in her forties who edits a magazine for sadists and masochists was revealing about the facts. First, she had no doubt that the pain she enjoyed had nothing to do with the pains of illness and injury, which she loathed. Next, the whipping that she invited had to stop short of serious injury, so the sadist who inflicted them had to be under her control and trusted. The pain was associated with a hugely increased sexual awareness. She felt her reactions to be like those of a horse "alerted and bounding after the thwack of a crop on its buttocks." She knew nothing in her background to explain her actions in any symbolic or associated terms.

I report this here without any of the sympathetic feeling that usually allows one to understand what is heard. Clearly this practice has many variations and extremes. Some can achieve sexual satisfaction only with pain. Some have defined requirements. The greatest pleasure over many years for a successful white banker was to be beaten up by a particular black woman. For some, the pain is self-inflicted during masturbation. Alfred C. Kinsey, author of *The Sexual Behavior of the Human Male* (1948), while masturbating would "insert an object into his urethra, tie a rope around his scrotum and tug hard on the rope as he maneuvered the object deeper." Some evidently search for increasing extremes and may kill themselves with asphyxia or wounding.

The word *pain* includes a reference to punishment, as in "on pain of death." Some achieve the tranquillity of absolution by punishing themselves for their sins or the sins of the world. The flagellants in Spanish religious processions appear exalted. A thousand years of Christian art portray candidate saints seeking wounds with radiant eagerness. The masochists are a small minority, but the rest of the community is fascinated by the paradox of their existence.

In this chapter we have seen that the public display of pain and the expression of private suffering are full of surprises. The amount of pain and the amount of injury are not tightly coupled. The time course of pain depends on the needs for escape followed by the needs best suited for treatment and recovery. The location of the pain may differ from the location of the damage. The public display of pain has the purpose of informing others of the patient's needs whereas the private suffering assesses the meaning and consequences of the patient's own miserable state. All pain includes an affective quality that depends on the circumstances of the injury and on the character of the victim.

The Philosophy of Pain

Before we explore the details of pain, we must consider the general plan of what we expect to find. The most common prevailing opinion, which comes from our intuition and is expressed by the majority of philosophers, is dualistic: that is to say, we have a body and a separate entity, the mind. The body is generally seen as a wonderful intricate machine operating on understandable principles that will be revealed by increasingly sophisticated scientific investigation. It includes a sensory nervous system whose function is to detect events in the world around us and within our own bodies. This sensory nervous system collects and collates all the available information and presents it in a form that generates pure sensation, according to the dualists. At this supposed frontier, the mind, which operates on entirely different principles, may inspect the sensory information and begin mental processes such as perception, affect, memory, self-awareness, and planning of action.

There are those who believe that these mental processes operate on such different principles that they will never be revealed by a continuation of present-day scientific exploration. Such people admit that the general outcome of mental processes can be described by psychologists but that the mechanism by which they are achieved will remain obscure. A more cautious group of dualists see mental processes as operating on principles that are entirely different from those of the body but that will eventually be understandable in materialist terms, including

17

obeying the laws of physics. However, we will take the approach that the abrupt frontier between body mechanisms and mental processes does not exist. Instead, I will propose that mind, body, and sensory systems exist as an integrated unity serving the biological needs of the individual with no abrupt shift of fundamental mechanism.

Dualism

If I tell someone I am concerned with the problem of pain, they frequently ask me whether I mean physical pain or mental pain. That question expresses the dualism of our culture. If I say, "My foot hurts me," I express the dualism of my thinking. One arm of my dualism is the thinking, talking, feeling, suffering, cognitive mental me. The other arm of my dualism is the stuff of my body, which sends a message to the mental one that it is damaged. The structure of the sentence is exactly the same as "The dial on my dashboard says that the engine is overheating." It all seems so transparently, intuitively obvious.

However, if we are to consider pain, we should first be quite certain of what appears obvious: that we inhabit two separate interconnected worlds, the physical and the mental. If dualism is that starting point, then the route of our exploration is set by an initial search in the machinery of the body for a sensory system that delivers messages that create sensation, after which there is a second phase of our search for the mental processing of received messages. That route has been taken for two thousand years from Aristotle to John Searle and Daniel Dennett. Pain has been used repeatedly as the simplest possible example of a physical stimulus that inevitably results in a mental response. We will not retrace this route, dropping the names of Bacon, Hume, Berkeley, Kant, and Wittgenstein, who have brilliantly described their version of the journey from sensation to perception. Nor will we join my fellow emeritus academics in their obsession to greet our oncoming senility with a discussion of consciousness. Two key philosophers will be sufficient for me.

Descartes wrote in 1640:

If for example fire comes near the foot, minute particles of this fire, which you know move at great velocity, have the power to set in

FIGURE 4. Descartes's picture of the pain pathway from *L'Homme* (1664). The fire pulled on "delicate threads" that opened pores in the commonsense center (F), which he located in the pineal gland.

motion the spot of skin on the foot which they touch, and by this means pulling on the delicate thread which is attached to the spot of the skin, they open up at the same instant the pore against which the delicate thread ends, just as by pulling on one end of a rope one makes to strike at the same instant a bell which hangs at the end.

This is precisely the same formal structure of a sensory signaling system that many accept today. Descartes was well aware that it was necessary to provide a transition zone between the bell and the next stage. He therefore proposed a further mechanism where the threads came together to reach the seat of imagination and common sensation. Here "external objects are able to impress the mind." He uses the word *esprit,* meaning mind, soul, or spirit.

In contemporary terms, Sir John Eccles, Nobel laureate neuroscientist, and Sir Karl Popper, philosopher, describe this area where the threads come together as the "liaison area." They state, "The unity of conscious experience is provided by the self-conscious mind and not by the neural machinery of the liaison areas of the cerebral hemispheres." They proceed:

> The self conscious mind can scan the activity of each module of the liaison brain or at least those modules tuned to its present interest. . . . The self-conscious mind has the function of integrating its selections from the immense patterned input it receives from the liaison brain in order to build its experiences from moment to moment.

Here contemporary scientists bring Descartes up to date and locate the frontier between body and mind, sensation and perception in the cerebral cortex.

Descartes was in trouble in his own day. A marquise challenged him to explain how his scheme permitted the phenomenon that a man with an amputated leg sensed in every detail his missing leg. This persistent observation by amputees, or "phantom limb," is not vague. An arm is felt in every detail with the hand and each finger. Descartes replied:

> In view of these considerations, it is manifest that, notwithstanding the goodness of God, the nature of man, in so far as it is a composite of mind and body, must sometimes be at fault and deceptive. For should some cause, not in the foot, but in another part of the nerves which extend from the foot to the brain, or even in the brain itself, give rise to the motion ordinarily excited when the foot is injuriously affected, pain will be felt just as though it were in the foot, and thus naturally the sense will be deceived: for since the same motion in the brain cannot but give rise in the mind always to the same sensation and since this sensation is much more frequently due to a cause that is injurious to the foot than by one acting in another quarter, it is reasonable that it should convey to the mind pain in the foot, rather than as in any other part.

In this single paragraph, Descartes had solved the superficial problem by inventing the concept of the false signal, which resurfaced three centuries later to plague electrical engineers. However, in so doing he had constructed a much more serious trap for himself. For him, the mind was passive and isolated, so utterly dependent on its input that it was unable to differentiate between true and false signals. While the basic dualist proposal that mind and body are separate entities may seem intuitively acceptable, the extended proposal that the mind is a passive dependent entity may seem less attractive. It leads to the concept of the cognitive self as an entity enclosed in a capsule and only able to read the dials on the walls of the capsule.

The phantom limb was not the only challenge. Spinoza, a younger contemporary of Descartes, wrote, "How can the cleverest man of our age descend into the very mysticism that he cleared away from the old academics?" It is true that Descartes was a revolutionary who had cleared classical philosophy from its old obscure waffle on the nature of cause and effect. Now, with his dualism, Descartes had separated an exactly described, definable sensory apparatus from a mystical mental miasma about which he had nothing to say. This abrupt transition from brilliance to gibberish needs exploration.

The most likely cause is to be found in the intellectual atmosphere of the early seventeenth century. The Roman Catholic Church remained in a dominant position and was clear about the human territory over which it had jurisdiction. A question to test the orthodoxy of a suspected heretic was about their acceptance of the transubstantiation of the host in the Mass, the conversion of the bread and wine into the body and blood of Christ.

It was exactly on this rock that Galileo foundered. Galileo was an older contemporary of Descartes. The reformation churches and thinkers rejected the reality of the transformation of the bread and wine. In this atmosphere, it is not surprising that Descartes declined to cross the frontier from the mechanism of the body into the mental territory owned by the Church. He most certainly wished to avoid a discussion of how faithful Christians attending Mass could observe and experience wine at one instant and blood at the next. Dualism was born partly because it seemed intuitively obvious that "the nature of man . . . is a composite of mind and body" and partly because it was inappropriate and heretical to treat mind and body as aspects of a single entity.

Descartes had written on his tomb stone *Bene qui latuit, bene vixit* ("He who hid well, lived well"). He certainly lived well, but from what was he hiding? He was certainly not hiding from what he called "the first principle of the Philosophy for which I am seeking." The principle was *Cogito ergo sum* ("I think therefore I am").

> From that I knew that I was a substance, the whole essence or nature of which is to think, and that for its existence there is no need of any place, nor does it depend on any material thing; so that this "me," that is to say, the soul by which I am what I am, is entirely distinct from body, and is even more easy to know than is the latter; and even if body were not, the soul would not cease to be what it is.

Descartes hides behind the authority of his own subjective introspection. He expresses complete conviction. How does he come by this certainty? Are there any facts? There are no facts. It is only a manner of thinking that seemed compellingly obvious and apparent to Descartes and to a very large number of people in his and our culture.

In order to separate the mind from the body, Descartes had to invent communication channels, the sensory nerves, by which external objects and internal events could "impress" the mind. He knew nothing of the details, so he had to invent them by guesswork. However, the general idea seems plausible if a mind-body separation is accepted. The sensory communication channels could be described in mechanical terms as linked chains and rods. Beyond this mechanism, the mind operated on fundamentally different principles that Descartes did not describe.

The alternative to dualism is monism, which proposes that mental processes are inherent outcomes of bodily processes. The power of the Roman Catholic Church to declare certain ways of thought to be heretical and thereby to dominate the course of philosophy may have declined. However, the growing exploration of monism as an alternative way for the analysis of thinking led to the proponents being accused of a new form of heresy. Monists are labeled simple-minded mechanists who deny any of the engaging properties of humans and who see the entire human being as nothing but a mass of interlocked gear wheels. Dualists of the Eccles and Popper variety describe a first-stage deterministic body machinery, rigidly performing its ordained tasks, that is observed by a conscious mind full of ideas, feelings, and emotions. By assigning all the glories of humans to an unobservable mind, they trivialize the body's machinery. The alternative is that the body is an integrated whole from whose properties emerge intellectually separable components.

For Descartes, the mind, soul, and spirit, and therefore cognition, were unique God-given properties of mankind and were not present in animals. In this respect, animals were considered to be God-created but then God-forsaken creatures placed under the dominion of humans. Animals were condemned to respond in a mindless mechanical fashion to circumstances. Obviously, this extreme separation of humans from animals has softened over the centuries. Respect for animal behavior has grown, not only because of Charles Darwin's introduction of animals as human relatives but because a study of their behavior reveals an unsuspected subtlety and complexity. Certainly animals, like humans, may appear to detect or to ignore injury, they exhibit very similar patterns of escape and recovery, and they learn to avoid harmful circumstances. They cannot tell us in words about the nature of their misery, but then, neither can babies.

Bertrand Russell was certainly a candidate for the cleverest man of our age. In his *History of Western Philosophy* he proposes a strict form of dualism, but for reasons quite different from Descartes. He begins at the top with rational thinking, which was the greatest challenge for him. His biography shows him rapidly bored with those colleagues, children, and discarded girlfriends incapable of rational thought as defined by him. His *Principia Mathematica*, written with A. N. Whitehead, was the achievement of many years of struggle to create a symbolic logic in which the rules of consistent relationships were laid out. The mind is evidently capable of manipulating abstract symbols and of placing them in logical categories. However, the ultimate origin of these internal mental digestions depended on observation of the nature of the external world at their beginning, and the verification of their rationality depended on checking with the outside world at their end. He was not concerned with hallucinations or fairy tales. The whole rational process begins with data (that which is given) from the outside world.

There is a political analogy very similar to this process. A dictator seeks to establish thought control by seizing all means of communication. When a civil war breaks out, the first target is the television network, followed by blocking the international telephone links and the expulsion of foreign correspondents. In this way, the dictator intends to feed the citizens only with the information he wants them to learn. A wise dictator, of whom there are fortunately few, takes great care to ensure that he is the one person in the country who has full access to all available information. All too often, the leader neglects this elementary precaution and surrounds himself with sycophants who tell him what he wants to hear and spin doctors who so skilfully manipulate the news that everyone is confused.

Russell understood that censorship of input would be a disaster and anathema to rational thinking. For this reason, he was more than content to support the Cartesian idea of a mechanical information center in which all the available information detected by the sense organs would be on constant uncensored display. He wished to assign to the mind the role of supreme commander fed by loyal, efficient, unquestioning messenger boys who would update him on every available bit of information, important or unimportant. This information service is entirely passive, whereas the mind is selective. Russell does not consider the role of exploration, where the organism actively seeks information. It is true that exploration has an inherent weakness because it implies a decision by the mind that one piece of information is more useful than another. However, as we shall see, every aspect of our conscious sensation suggests that it is not fed by a passive machine but includes brain activity, which has directed attention, and ordered exploration, which has selected part of the sensory input and has amplified information about details. Unlike Russell's ideal of a passive input analyzed by an active brain, there are signs that brain activity controls the input. This does not mean that the entire outside world is a hallucination, but it does mean that our senses include active participation of mind and body.

If we define the mind as those processes within us about which we can give verbal reports, we should begin by examining processes about which we have no cognitive knowledge.

Behavior Without Mind

All of us, like Russell, are hugely impressed by the mind. Since we are here to consider pain, we should define precisely the point at which the mind is aware of pain. Are there important processes that can precede the mental awareness of pain? Because most of us are champions of the mind and sing its praises, "intellectuals" tend to reserve a series of derogatory words for body processes that do not penetrate the mind, calling them automatic, mechanical, reflex, or instinctive. Quiet breathing is just such an event, proceeding for hours and days accompanied by mental ignorance. Yet, if a drop of fluid is inhaled, the mind pounces into action. Furthermore, if you wish to blow your trumpet, you override your normal breathing with skilled control. Breathing is only an automatic mechanical reflex when operating in preset limits. However, the mind is involved as soon as the limits are exceeded. How were those limits set? By genetics? By learning? The threshold for switching from automatic to mental is variable and under control.

Many of us have had the experience of driving a car while in conversation with a friend or in silent thought and have suddenly become aware that miles have gone by and we have no recall of the route but remember the conversation perfectly. Was the driving automatic? If by "automatic" one means the execution of a preset program of motor movements, the answer is no, this is not automatic driving. In the course of the journey, the driver had to react to stop lights that could not have been preprogramed by the driver. Therefore, this "mindless" episode involved the reception and analysis of sensory signals as well as the execution of motor orders.

The greatest exponents of this feat are professional athletes. They repeatedly report that their highest achievements were attained at a time when they had entered "the zone." In the zone, they report that the mind switched off from all detail and they experience a feeling of exhilaration and freedom. In the case of a 100-meter sprinter, one could propose that the entire race had been preprogramed during training so that, from the starting line to the finishing tape, a repeated pattern of motor orders were issued to the muscles and the sensory system was completely switched off. This would then be the achievement of an automaton. For those of us who are inactive slobs, it may comfort us to downgrade athletic achievement by downgrading the athlete to robot.

However, this explanation could certainly not apply to champion tennis players, who are obviously reacting to the sensory challenge presented by the flight path and spin of the ball as well as generating precisely appropriate motor responses. This is not an intellectual achievement because no thoughtful cognition was involved at any time. When the athlete watches his performance on video, he does not relive the experience. He watches himself for the first time.

It is true that the most impressive of these mindless behaviors usually involve complex motor movements. In order to preserve intact the glories of the cognitive speaking mind, it has become common to assign them to a special category: motor learning. It is true that professional pianists describe how they perform a sonata while thinking at most of a broad general structure and certainly not note by note. It is also true that, in their infancy, those same pianists did think and play note by note. The transition from unskilled to skilled movement is accompanied by this ability to stop thinking from point to point and to achieve a whole. I do not believe that we can put these astonishing achievements to one side by labeling them motor learning.

The chess grandmaster Gary Kasparov will play twenty gifted amateurs at a time and beat most of them, walking from one board to another and taking only seconds to make his move. He reports that he makes his mind blank during these multiple games. I take this as a superb intellectual assessment of the patterns of the pieces, although the actual motor movement itself is trivial and could be carried out without skill. When the sensation of pain engulfs the mind, elaborate processes of analysis of which we are not aware may accompany the appearance of pain in cognitive awareness.

The Senses

Aristotle said that there were five senses: sight, sound, smell, taste, and the body sense. By this he meant more than the obvious fact that there were five sets of organs responsible for the five senses. Each sense was a unique category in itself that could not be imitated by any mixture of the other four senses. It is true that the blind poet Milton wrote of "the scarlet sound of the trumpets," but that refers to association "synesthesia" not to imitation. With human obsession with subdivision, these huge categories of sensation could not be left long to stand by themselves. Color was itself subdivided into its primaries in such a way that mixtures of the three primaries could generate all the colors.

For skin senses, it was decided by introspection that there seemed to be four or six primary sensations: touch, warm, cold, pain, tickle, and itch. As with color, it was proposed that all other sensations were composed of mixtures of primaries so that, for example, wet was a mixture of touch and cold. Loud alarm bells should sound in anyone's brain on reading a list like the one above. It mixes up classes of words but invites you to treat them as the same class. Touch, warm, and cold describe the stimulus; pain, tickle, and itch describe the mental response, the sensation. In nineteenth-century terms, the sensation seemed so closely linked to the stimulus that it did not matter much which class of words were used, for light pressure seemed to equal touch and heavy pressure to the point of damage seemed to equal pain. The major lesson of chapter 1 was that tissue damage and pain are not so intimately linked that the two can be considered equivalent. We must therefore be very cautious and use one set of words for a stimulus event and another set for a perceived sensory event.

The Sense of Pain

Aware that they could no longer accept Cartesian thinking, which insisted that a sensation was no more than a mental representation of a stimulus, the International Association for the Study of Pain asked a group chaired by the psychiatrist Harold Merskey to provide a modern definition of pain. They said: "Pain is an unpleasant sensory and

SENSORY

	Temporal	Spatial	Punctate Pressure	Incisive Pressure	Constrictive Pressure	Traction Pressure	Thermal	Brightness	
	FLIKERING		PRICKING		PINCHING			TINGLING	1
	QUIVERING	JUMPING	BORING		PRESSING	TUGGING	HOT	ITCHY	2
	PULSING	FLASHING			GNAWING	PULLING		SMARTING	
	THROBBING	SHOOTING	DRILLING	SHARP	CRAMPING			STINGING	3
	BEATING		STABBING	CUTTING		WRENCHING	BURNING		
	POUNDING		LANCINATING	LACERATING	CRUSHING		SCALDING		4
							SEARING		5

SENSORY — AFFECTIVE

	Dullness	Sensory: Misc.	Tension	Autonomic	Fear	Punishment	Affective: Misc.	
		TENDER					Anchor words	1
	DULL							2
	SORE	TAUT	TIRING			PUNISHING		3
	HURTING	RASPING	EXHAUSTING	SICKENING	FEARFUL	GRUELLING	WRETCHED	
	ACHING				FRIGHTFUL		BLINDING	4
	HEAVY	SPLITTING		SUFFOCATING	TERRIFYING	CRUEL		
						VICIOUS		5
						KILLING		

EVALUATIVE

	Anchor words	
	MILD	1
	DISCOMFORTING	2
	ANNOYING	
	DISTRESSING	TROUBLESOME 3
		MISERABLE
	HORRIBLE	INTENSE 4
	EXCRUTIATING	UNBEARABLE 5

emotional experience associated with actual or potential tissue damage or described in terms of such damage." They added some crucial notes: "Pain is always subjective. . . . This definition avoids tying pain to the stimulus."

In the classical view, stimulus leads to pure sensation, which leads to perception. Tissue damage leads to pure pain, which leads to pain and unpleasantness. But in the modern view, stimulus may or may not lead to perception. Tissue damage may or may not lead to pain, which is an unpleasant experience.

There were two reasons for the new definition. First, observation and introspection failed to find a correspondence between tissue damage and pain. Second, no ordinary person ever experienced a pure pain that was not accompanied by unpleasantness. It is an honest definition of what we observe in ourselves and others, and it avoids the step invented by neat-minded philosophers that there must be at first a pure sensation that is later followed by the mental value judgment, which assigned emotional value, such as unpleasantness, to the pure sensation.

You may well question the usefulness of using only the single word *pain.* To bring order to our world, we humans are passionate classifiers. We love the act of sorting and assigning pigeonholes. Now we have a definition of pain. Pains can be big or small, but is it true that one toothache equals two headaches? Melzack spent many years collecting words that people used to describe their pains. He found seventy commonly used words, which he sorted into categories. Some words, such as *pricking* or *hot,* seemed to be used just to describe the stimulus itself. For each of the classes of sensory word, the words were arranged in order of intensity; for example, *hot, burning, scalding, searing.* Then there was another class of words that he called affective, which described what the sensation was doing to the victim; for example,

FIGURE 5. The words that Ronald Melzack heard patients use to describe their pain. He divided them into three large groups: sensory, affective, and evaluative. For each group, he ordered the words in term of implied intensity, with weaker words above and stronger words below. They formed the basis of the McGill pain questionnaire (*Pain* 1, 277–299; 1975).

exhausting, sickening, punishing. Finally, he separated out words that he called evaluative, which expressed the degree of suffering; for example, *annoying, miserable, unbearable.*

Patients were presented with the full list of words arranged in boxes, and they were asked to tick which words best described their pain. Try it yourself. You may disagree with the way he has arranged the words. Rearrange them. Add your own if you think words are missing. Very extensive testing of this method, which became known as the McGill pain questionnaire, showed patients falling into loose groups that were characteristic of their disease. We will return to their answers. One example has already been discussed for patients entering an emergency room where all patients were eventually in pain. They used words about their pain that did much more than describe its intensity. The word *pain* for each person had at least three dimensions beyond its intensity. Melzack called the dimensions sensory, affective, and evaluative. You might wish to describe your pain in other ways. The point here is that the single word *pain* has several components that combine in the individual to express what they mean by their pain.

In this chapter, we have examined and criticized the proposition that the mind and body are separate entities. The individual is certainly capable of an elaborate analysis of events and of the generation of skilled responses, but these may occur with no cognitive awareness. Pure pain is never detected as an isolated sensation. Pain is always accompanied by emotion and meaning so that each pain is unique to the individual. The word *pain* is used to group together a class of combined sensory-emotional events. The class contains many different types of pain, each of which is a personal, unique experience for the person who suffers.

The Body Detects, the Brain Reacts

In the first two chapters, we dealt with the whole person in pain, but now it is time to turn to the details of what precisely happens when pain is provoked, so that we can later bring these details together to understand the whole. You may not wish to struggle with the unfamiliar territory of this chapter because it deals with the detail of how the nervous system handles the news of unexpected events. If you wish, skip to chapter 4, where I begin to describe the experience of pain as a whole. However, you may wish to return to this chapter later because the overall experience of pain emerges from the coordinated action of the parts. Damage to tissue initiates an operatic drama in three acts of action and reaction in the tissue. The dramatis personae are the sensory nerve fibers, cells in the damaged tissue, blood vessels and their contents, and the sympathetic nervous system.

Sensory Nerve Fibers

Sensory nerve fibers originate from clusters of cells that lie close to the spine, with one cluster or "ganglion" for each vertebra. A special ganglion lies in the base of the skull and supplies the face, mouth, and head. In the embryo, each cell puts out a short fiber that splits at a T-junction. One arm grows out into the tissue by way of the nerves. The other arm grows into the spinal cord with a large group of similar nerve fibers called the dorsal root, which contains all the fibers from the ganglion. The skin is profusely innervated with three types of sensory fibers. One group, called A beta fibers, are wrapped in a fatty protein called myelin and are sensitive to gentle pressure. The second group, called A delta fibers, are thinner and are sensitive to heavy pressure and temperature. The third group, called C fibers, are very thin and have no myelin and respond to pressure, chemicals, and temperature. Deep tissue and organs such as the heart, bladder, and gut are innervated only by the thinner fibers.

Sensory nerve fibers detect events occurring at their ends in the tissue and signal them to the spinal cord by two methods. One method is by the production of nerve impulses, which are abrupt events in the membrane of the fiber that sweep over the fiber from its peripheral end all the way to its central end in the spinal cord. These impulses travel along the fiber like fire in a fuse but, unlike a fuse, the surface of the fiber rebuilds itself after the fire has passed. The impulse lasts only one thousandth of a second and travels at between 1 and 100 meters per second depending on the thickness of the fiber. Some of our sensory nerve fibers are more than a meter long, running from the toes to the middle of the back; others are only a few centimeters long, running from the teeth to the hindbrain.

The second method of sensory nerve communication is much slower. The ends of nerve fibers in the tissue, particularly the thin C fibers, absorb chemicals from the tissue and slowly transport these chemicals all the way to the cell bodies in the ganglia and on to the central terminals. In this way, the sensory nerves have two ways of informing the central nervous system about changes in the tissue in which the sensory nerves terminate. One way is by volleys of nerve impulses that bombard the central cells on which they end. The other way is by the slow transport of chemicals, which change the action of cell bodies in the ganglia and the central terminals. This transported signal takes a few hours to reach the ganglion cells if the injury is very close and many days if the damage is as distant as the foot.

Cells in the Damaged Tissue

Each part of the body is made up of a collection of different types of cell. Some of these are special to that region, be it skin, heart, or muscle, and others are common to all types of tissue, such as the cells that make tendons and fat. Some of these cells are very fragile, whereas others are tough, resisting anything short of devastation. A mild skin burn leaves the general structure intact while some cells have fallen to bits. Obviously, no cell can withstand a direct hit from a bullet.

Blood

Red blood cells in tissue are the most dramatic and least important aspects of injury, assuming that the bleeding stops. The color of a black eye or a bruise is caused by the hemoglobin in blood that has leaked into tissue, and the gaudy rainbow of colors that follows is a sign of the breakdown of the haemoglobin. Far more important are the white cells that migrate from blood vessels into damaged tissue even if the vessels are not cut open. They set about engulfing the debris of damage and recognize and counteract the presence of abnormal components by way of the immune system. This invasion occurs in all sick tissue but is particularly dramatic if bacteria have started to multiply in the damaged tissue when the white cells form a barrier to create a pus-filled abscess. The fluid of the blood, called the serum, also plays a crucial role. Elaborate interaction of proteins in the serum with damaged cells forms the clot that blocks leaking blood vessels. Other active components of serum move into damaged tissue and are evident in the swelling of affected tissue.

The Sympathetic Nervous System

In peacetime, these fibers regulate the amount of blood flowing through tissue and regulate the steady activity in such organs as heart, gut, bladder, and sweat glands. In emergencies, the fibers change their action when their cell bodies are informed of damage by a reflex loop originating in the increased activity of the sensory nerve fibers. Apart from regulating blood flow, they play a role in triggering some of the inflammation that is characteristic of tissue damage.

Tissue Damage

Following tissue damage, a three-act opera is played out in the form of immediate, secondary, and tertiary responses.

Immediate

When tissue damage occurs suddenly, there is no overture, although more commonly there is slow accumulation of abnormality until the effects are overt. Three changes can immediately provoke pain by way of impulse activity in the sensory nerve fibers. Pressure is one obvious cause, and it is obvious that sudden pressure, as in a slap, is more effective than slow-onset pressure. Heat and cold may trigger pain, which is signaled by impulse activity in fine sensory nerve fibers. Chemicals may seem a less obvious cause of pain, but we have all experienced the effect. Mustard, curry, and chili, if overdone, produce burning pain. Insect bites and nettle stings inject pain-producing substances under the skin. The CS spray used by the police produces pain where it is easily absorbed in the eyes, nose, mouth, and skin. Much more important, the breakdown of cells following damage produces an array of pain-producing substances.

Secondary

When the fine sensory nerve fibers are stimulated, they emit into the tissue substances called peptides, which in turn cause dilatation of the blood vessels and pain. When cells are smashed, their contents spill out into the tissue, and some of these chemicals cause pain. With the disruption of cells, compounds not normally present find their way into contact with enzymes, biological catalysts that break down the debris into smaller molecules, some of which produce pain.

Tertiary

The classical cardinal signs of inflammation were tumor, rubor, calor, and dolor (swelling, redness, heat, and pain). Now we can describe in detail what is happening. The swelling is produced by the leaking of fluid into the tissue from the dilated blood vessels. In addition to fluid, there is a massive invasion of white cells attracted from the blood by substances released from the broken cells and their breakdown products. The redness is simply due to the widening of the usually narrow blood vessels. The heat is produced by the large increase of hot blood now rushing through the dilated blood vessels. The pain is produced

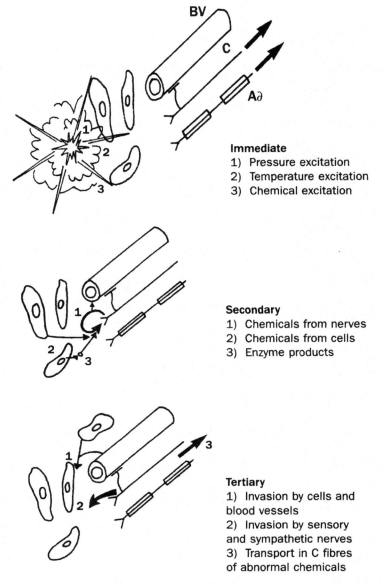

BV

C

A∂

Immediate
1) Pressure excitation
2) Temperature excitation
3) Chemical excitation

Secondary
1) Chemicals from nerves
2) Chemicals from cells
3) Enzyme products

Tertiary
1) Invasion by cells and blood vessels
2) Invasion by sensory and sympathetic nerves
3) Transport in C fibres of abnormal chemicals

FIGURE 6. The three stages of change in tissue following damage. The tissue contains cells, blood vessels (BV), and two types of sensory nerve fiber that detect damage: A delta fibers, which are small fibers wrapped in a fatty sheath, and C fibers, which are very small naked nerve fibers.

by direct action of the original destructive event and is amplified by the soup of pain-producing chemicals made by the breakdown products. The area is also tender: that is, pain is now produced by gentle pressure that would previously have caused only the sensation of touch. The tenderness comes partly from the sensitization of the nerve ends, as they are soaked in pain producers, and partly by changes in the central nervous system, which we will discuss later. Some C fibers become active only in the presence of inflammation; otherwise, they are sleeping.

The finale of the opera occurs as these reactive processes reach their climax, once dead cells have been swept away by the white cells, and reparative processes begin. Cells capable of producing dense, fibrous tissue, called fibroblasts, move in and pack together to produce the familiar white scar. They may be accompanied by growing blood vessels and nerve fibers to produce the tender, angry red scar. During all this time, the C fibers have been absorbing the unusual chemicals present in the inflamed tissue and transporting centrally them to change central excitability.

Nerve Damage

Severe injury, of which amputation is an extreme example, may cut across whole nerves. There are also medical causes of nerve damage, such as the invasion of nerves by viruses, which produces shingles, or the metabolic failures in diabetes that cause a breakdown of nerve fibers. These conditions produce all the standard signs of inflammation we have just described but, in addition, are particularly likely to produce pain that outlasts inflammation. The cut nerve attempts to regenerate and sends out fine sprouts that have unstable properties, especially if the regeneration fails to send the fibers back to their former target. In addition, the cut nerve absorbs highly unusual chemicals that are transported back to the ganglion cells and to the spinal cord, where they trigger exaggerated sensitivity, as we will discuss later.

The Spinal Cord

The sensory nerve fibers stream towards the spinal cord, gather together in the dorsal root, and enter the spinal cord (arrow entering the cord in figure 7). The special nerve for the head and face, called the trigeminal nerve, has exactly the same organization, but its fibers enter the hindbrain, which is the forward extension of the spinal cord.

The spinal cord (figure 7) is a cylinder of nerve tissue running the entire length of the body within the vertebrae. It enters the head through a large hole in the base of the skull and continues as the brainstem. In cross-section, the cord is round with its outer parts, shown in black in the photograph (figure 7), made up of a mass of central nerve fibers, which connect the cord to the brain, and of nerve fibers running in the opposite direction, which connect the brain to the spinal cord. Within this nerve fiber highway, there is a clear area containing the nerve cells. The upper area, called the dorsal horn because it is near the back (dorsum), contains the cells that receive the sensory messages, process them, and send on the analyzed messages. The lower area, called the ventral horn because it is near the belly (ventrum), contains the cells that generate the motor orders and send their messages back to the muscles and tissues (arrow leaving the cord in photograph figure 7).

We are concerned with the upper sensory part. The cells are arranged in laminae that run the entire length of cord. The upper three laminae make up a clear area (laminae I, II, and III, seen in figure 7). Here all the incoming small C fibers end on cells that also receive inputs from the other sensory fibers. The cells in this area have a wide variety of shapes, seen in the lower part of figure 7. They are mainly concerned with local, important, busy work, which we will describe. Only a few of them send messages to the brain. The lower laminae (IV, V, and VI, seen in figure 7) contain larger cells that receive the larger sensory input fibers (A beta and delta) and send their messages on to the motor ventral horn and to the brain. All parts of the cord receive sensory messages and send on processed messages to the motor part of the cord and to the brain, but it is essential to realize that none of this mechanism works properly if it does not receive messages from the brain. These modulating descending controls reach each area of the cord by way of fibers running in the fiber tracts in the outer rim of the cord.

Let us step aside for a moment to examine two cases that illustrate the relation between the brain and the spinal cord.

Case 1

A young Irish soldier was shot in the middle of his back while he was in a United Nations peace-keeping unit. A single bullet had cut completely across his spinal cord and he was immediately paraplegic. When he was examined some months later, his legs lay floppy on the bed and he could will no movement at all in them. He could feel no sign of any stimulus applied to his body below his navel. Yet the part of his spinal cord below the lesion was working in its fashion, even though it was isolated from the brain, unable to send or receive messages to or from the brain. Tapping his knee produced a lively jerk. Running a thumbnail along the bottom of his foot produced a powerful, long-lasting withdrawal of the leg, which relaxed after many seconds with a long series of jerks. Even his bladder emptied automatically without his knowing when it filled up. Yet he sensed a clear phantom image of his body with his legs in a knees-up position, even though his legs were in fact straight. A grim, deep pain was developing in his completely numb body.

Case 2

A lively, intelligent young woman had suffered localized brain damage during her birth. Her right arm was held, unmoving and flexed with

FIGURE 7. *Top:* the spinal cord in cross-section. The spinal cord is cylindrical and lies within the vertebrae. In cross-section it is round with long-running nerve fiber tracts on its outer edge, shown here in black. Sensory neurons enter at "IN" over dorsal roots (*dorsal* means "toward the back"). Motor neurons exit at "OUT" and form the ventral roots (*ventral* means "toward the belly"). In the center of the cord is the gray matter, where the nerve cells reside and where sensory information is collected, analyzed, and dispatched to the brain and to the motor cells in the ventral gray matter. The cells are organized in laminae, and in the dorsal horn of the grey matter there are six laminae, one above the other, labeled I-VI.

Bottom: cells in the upper five laminae of the dorsal horn. Each nerve cell has a rounded body containing the nucleus and the main metabolic machinery. From the cell body, elaborate treelike branches, called dendrites, spread out to collect the incoming information. Each cell body also sends out a nerve fiber, called the axon, which transmits nerve impulses to other cells. These cells were drawn from human cord by the Belgian neurologist, Jean Schoenen.

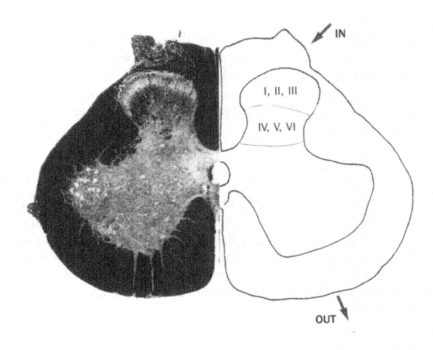

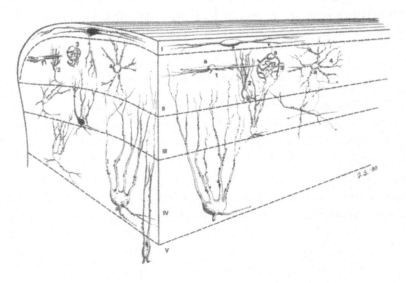

her fingers curled. She could move it a little in a weak, clumsy fashion. She could not send motor orders to the right arm area of her spinal cord. She could sense touch and pain in the arm but in a peculiar way. It ached all the time. She could not tell if a touch was moving or stationary. If a number was drawn on her arm, she had no idea what was happening. She was unable to send modulating orders down to the cord, which would have permitted her to unravel subtle aspects of the sensory messages.

Handling Sensory Messages

Let us take as an example one of the large cells in the dorsal horn in laminae V that responds to sensory messages and sends its order to the brain and to motor cells. The cell is represented as the large open circle in figure 8 with its input from the large sensory input fibers, A beta, and from the smaller sensory fibers, A delta and C. No cell is ever found that simply relays the inputs onto the output. There are always small surrounding cells that modulate the action of the input on the output. Some of these small cells, shown in black, exaggerate the effects of the input; others, shown as open circles, diminish the effect. All of these cells are also influenced by fibers descending from the brain.

Immediate Gate Control

If the input message comes in only from the large A beta fibers as a result of touch, the cell fires briefly and then is turned off by the action of the small white cell (upper part of figure 8). If, however, the input volley comes from the tissue damage detection fibers, A delta and C, the cell fires more vigorously and the small black cells also come into action and exaggerate the output. During all this time, the brain is sending down control messages to amplify, diminish, or ignore the signal, reflecting the arrival and reception of a sensory message.

Let us examine the action of immediate gate control in real situations. If you hit your thumb with a hammer, it hurts. But then you do many strange things. You wave it about, grip it, rub it, and run cold

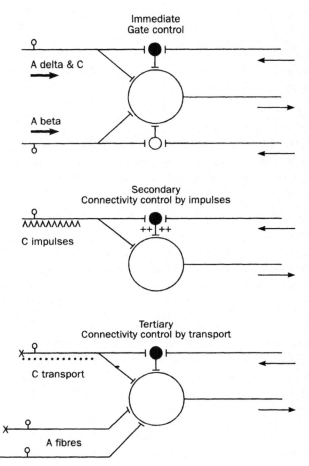

FIGURE 8. The three stages in which connectivity of cells in the dorsal horn of the spinal cord is modified. These cells receive incoming sensory messages, which announce the presence of injury. The injury messages arrive over the A delta and C nerve fibers shown in the tissue in figure 6. They excite cells in the dorsal horn, shown as a large white circle. Surrounding this cell are small excitatory cells, shown in black, and small inhibitory cells shown as small white circles. The output of the large sensory cell is shown leading to the right and terminates on motor systems and in the brain. The brain sends down controlling systems, shown as the axons arriving from the right. These neural circuits allow the output to be controlled.

water on it. You are stimulating large, low-threshold A beta fibers, which in turn are stimulating the small white cells that diminish the firing of the big white cells. This is the basis of rubbing, scratching, and massage for pain. When William Sweet, chief of neurosurgery at Harvard, and I first realized that this was the effect of the large fibers, we took advantage of the coincidental fact that such fibers can be stimulated very easily by electricity and we invented transcutaneous nerve stimulation (TENS), to be described later.

If you are about to give blood and you watch carefully as the rather large needle approaches your skin and penetrates the vein with a sudden rush of dark blood into the syringe, you feel a sharp, stinging, tearing pain that can be quite alarming. On the other hand, if you concentrate on some handsome nurse who is wandering around the room or, even better, if you close your eyes and snuggle down in the chair, fantasizing that you are on a beach warmed by the sun and listening to the waves, you hardly feel anything beyond the small prick. This is because the brain has sent messages down to the cord that it has more interesting matters to which it has decided to attend. We know that this involves the spinal cord because the effect is not just a matter of sensation, as the local muscle reflexes are correspondingly increased or diminished. We know from the example of the paraplegic soldier that these reflexes are the business of the spinal cord, even though the brain can influence them. Similarly, if the nurse squeezes or slaps the skin before the needle enters, you are distracted and may not feel the needle.

Secondary Connectivity Control

If the input volleys contain a substantial contribution from the C fibers, the cells fire as we have just described but then long-term secondary excitabilities set in. The heavy bombardment, aided by the emission of peptides from the C fibers, sets off a cascade of chemical changes. The big white transmitting cell becomes more excitable and goes on firing after the input has dropped.

If you slip off a curb and badly twist an ankle, you are likely to feel two quite different pains. First you feel a sharp, quick, intense pain in the ankle that fades in seconds. Then a new type of pain may start up.

It is deep, spreading, sickening. Slowly the ankle becomes tender and the tenderness spreads to the whole foot and lower leg. You don't like anyone touching it and you hobble. The first phase of the pain involved the immediate gate control mechanism that permitted the impulses to enter the cord and set the amplification. The second phase involved the slow changes of excitability and connectivity of cells in the region that received the massive input volley. Patients who have been operated on feel only the second phase because they are anesthetized during the first. In an emergency, neither pain is felt until the emergency is over, as we have seen in chapter 1.

Tertiary Connectivity Control

When nerves have been cut across or when inflammation persists in the tissues, abnormal chemicals are transported to the cord by C fibers. These exaggerate the excitability of the spinal cord transmitting system. Worse, they immobilize the turn-off mechanism in the spinal cord, represented by the small white circle cells. Small inputs result in large outputs. In the extreme, these hyperexcitable cells begin firing by themselves, sometimes in bursts and sometimes steadily.

The amputees described in chapter 1 showed all of these changes. All of them had patches of skin on their stumps that were exquisitely sensitive, even though the sensitive tissue had never been injured. All of them had clearly perceived phantom limbs, some of which were felt as continuous pain described as hot, burning, or cramping. Some had stabs of pain that seemed to penetrate the whole missing limb and shook the whole body so that one could witness the man's reaction by his blanching of the face, gripping, and tensing.

Destination of Nerve Fibers

The spinal cord as it enters the skull merges with the medulla, an opened-up version of the cord containing sensory and motor areas. Further forward, the medulla becomes the pons, on top of which the cerebellum resides (figure 9). Further still is the midbrain, which again closes up to become a cordlike cylinder with sensory areas on top and

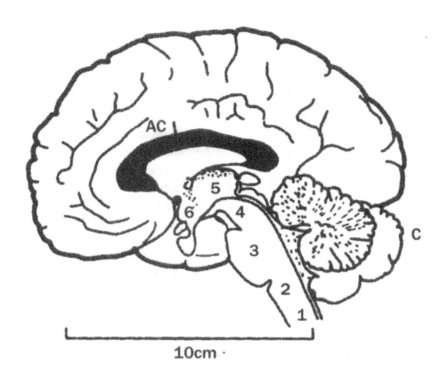

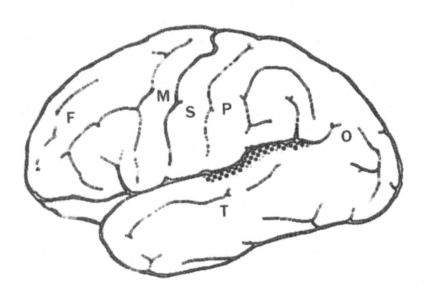

motor areas below. This brainstem made up of medulla, pons, and midbrain receives ascending messages from the cord and a special input from the viscera and adds inputs from eyes, ears, and face, mouth and head. In addition, it distributes orders to eye muscles, face, and jaws.

Finally, beyond the brainstem is the forebrain. At its core is a mass of cells, the thalamus and basal ganglia. Below the thalamus is the hypothalamus, which is concerned with all our internal business: feeding, blood flow, temperature, hormones, and so on. Wrapped completely around the thalamus is the cortex. It consists of a vast, folded, continuous sheet of cells in six layers that receive inputs from everywhere, send impulses everywhere, and, for good measure, talk to each other.

Every one of the structures mentioned receives powerful input signals generated by tissue damage and coincident with the production of pain. In the rest of this book, we have to unravel this apparent mess. No one area has the monopoly of capturing the one and only input signal associated with pain. One thing is certain: we are not going to find a single pain center as proposed by Descartes. At the same time, this huge accumulation of nerve cells is not a random network because it contains specialized zones that provide some hope for the discovery of an overall plan. It is obviously time to put our thinking caps on top of our brains in order to propose an acceptable subtle theory that takes the anatomical and physiological facts into account.

One fact we have already stressed is that the brain does not sit passively reading the sensory messages sent to it from the tissues and spinal cord. It sends out descending control systems that shape the

FIGURE 9. *Top:* the human brain sliced down the middle. The spinal cord (1) enters the skull and merges with the medulla (2). The brain stem enlarges to become the pons (3), on top of which lies the cerebellum (C). The brain stem continues as the midbrain (4), which terminates in the forebrain, which contains the thalamus (5) and the hypothalamus (6). The rest of the figure is the cerebral cortex, which balloons out from the forebrain. One part of the cortex is the anterior cingulate cortex (AC), which we mention in the text.
Bottom: side view of the human cortex showing its four lobes: frontal (F), parietal (P), occipital (O), and temporal (T). The classical motor area (M) and the classical body sensory area (S) are also shown.

received messages. Most of these descending control systems originate from cells in the core of the brain stem, in the medulla, pons, and midbrain. These cell groups are in turn affected by inputs from the cord and from the forebrain. We are therefore faced with a feedback in which the input is affected by the brain as well as by events in the periphery. The most famous descending control runs all the way from cortex to cord. It has therefore been assumed to be the motor system by which the brain orders the muscles to contract. It now turns out that even this corticospinal pyramidal motor system also influences the sensory message received by the cortex.

In summary, this chapter has outlined several stages. When tissue is damaged, a sequence of events produces inflammation with pain. The spinal cord is informed of tissue damage by way of sensory nerves. Cells in the spinal cord react immediately to the input, but the amount of their output depends on small cells that can enhance or diminish the output message. The setting of the small cells depends on orders descending from the brain. Once bombarded by injury messages, spinal cord cells exaggerate their sensitivity. Chemical messages from damaged tissue, and especially from cut nerves, further increase the sensitivity of cord cells. The cord cells signal to many parts of the brain that injury exists. Many parts of the brain then feed back onto the cord cells and amplify or reduce their output messages.

4

The Whole Body

When thinking about pain, we naturally concentrate on our conscious awareness and tend to ignore all other associated events. We use put-down words such as *reflex, automatic,* and *mechanical* to distinguish them from the main event that dominates us: the conscious experience of pain. We should be cautious at this stage of our search. If pain is a puzzle, we should not throw away pieces of the jigsaw just because we are obsessed with a preconceived single solution.

William James, brother of the author Henry, was a brilliant professor of psychology at Harvard at the end of the nineteenth century. He proposed that some of our emotions were an awareness of our general body reactions to an event, rather than to the event itself. Fear, for him, was an awareness of sweaty palms, a dry mouth, trembling, and a pounding heart. Many dismiss this approach as a Victorian mechanical attitude to the mind because they wish to preserve the mind as a self-contained independent entity. We should, however, examine the possibility that pain is a syndrome that joins together a coincident group of signs and symptoms rather than a single phenomenon. It could be that we are aware of the combination of events rather than of pain as a single separate event. Some of these associated reactions have the huge advantage that we can detect them in other people, adults and babies, and in animals that have no recourse to verbal communication.

Alerting and Startle

We observe this when something new happens in the world that could be important but that has yet to be identified or located. Each species is marked by its special sign. The blackbird in the tree shouts a series of brief chirps. Chicken chuckle. Deer run. The dog freezes in its tracks, one foot lifted, tense, trembling, and silent. Humans show their alertness by a posture that is the opposite of relaxation. If the novel event is large, we produce a startle response. The eyes open wide and the head flicks back, the arms flex and the legs extend. This startle response is apparent in the newborn baby and is used as a diagnostic test of how much of the brain is working. A gun unexpectedly fired behind you produces one hell of a startle. If the gun goes on firing, the startle drops with each explosion because it is no longer a novel event, and the startle is replaced with a steady alert tension that persists long after the gun stops firing, sometimes even for a lifetime.

Orientation and Exploration

Once alerting has occurred, there follows behavior that attempts to locate and identify the stimulus. Even Descartes showed the victim's face and eyes turning toward the stimulus (see figure 4). He reasonably interpreted this orientation as aligning the sense organs of eyes, ears, and nose to maximize the collection of as much data as possible about the object that has triggered alerting. These movement patterns are also species specific. A rat with a small skin wound sniffs at it and bites. A dog sniffs, licks, and scratches. We and our monkey relatives look and then probe with our fingers. All of this goes beyond simply locating the problem. We carry out active movements to explore the site of the stimulus and to collect all the sensory data possible to identify the nature of the stimulus. While this is going on, all other movement is interrupted.

Attention

These movement patterns are the outward signs of attention. We know in ourselves that attention goes beyond identification because it merges with an assessment of meaning and consequences. Attention is an integral part of pain. Pain captures and monopolizes attention and includes an interruption of any activities not directly related to pain relief. Many therapies, as we will discuss, attempt to recapture the attention from its imperial domination by pain. "Mommy will kiss it better" is a highly effective pain therapy that unfortunately fades as doubts rise about mommy's omnipotence. For adults, we have to invent more and more elaborate methods of distraction to liberate the brain from its pain master. In an emergency, the attention may be occupied by high-priority demands, such as escape, and pain may not be permitted to attract attention. In these situations, pain does not occur, as we saw in many examples in humans and animals in chapter 1.

Muscle Responses

Just about every high school biology text contains a diagram where a finger touches a saucepan and is rapidly withdrawn. It is used to "explain" pain as the method of avoiding injury run by a reflex mechanism consisting of sensory afferents that make motor nerves withdraw the hand. I despise that diagram for its triviality. I would estimate that we spend a few seconds in an entire lifetime successfully withdrawing from a threatening stimulus. Unfortunately, we are destined to spend days and months in pain during our lifetime, none of which is explained by that silly diagram. It is true that we and all creatures rapidly withdraw from a threatening stimulus by using a simple input-output loop of nerve fibres. Even the eye blinking is a special example of such a reflex. The brain is not involved in the existence of this reflex but is very much involved in controlling its size. In a paraplegic with the spinal cord completely cut across, a gentle tweak to a toe provokes a particularly violent withdrawal, even though the person feels nothing and cannot willingly withdraw the leg.

Let us pause here to consider the fact that, very, very rarely, children are born who grow up with no sensation of pain. This is called congenital analgesia, and these children are completely normal in other respects. We are not considering here those poor children who are born with very severe general central nervous system defects and seem not to notice a pinch. The otherwise normal children with congenital analgesia have been the subject of intense study because they are so fascinating and test all our ideas about the meaning and usefulness of our normal ability to perceive pain.

One such case was studied over a period of twenty years up to the time she was a student at McGill University in Montreal. A strong pinch to the foot failed to produce a withdrawal or to provoke pain. When pinched and asked what she felt, she replied in a calm way that "it feels like a very strong pressure, and I know that if you pinch much harder you will injure my foot." All her other body sensations-touch, pressure, warm, cold, and movement-appeared completely normal. How had she grown up without the massive protection supposedly provided by the withdrawal reflex? She had continuous monitoring by her doctor father, mother, and siblings, who were all aware of her problem. Gross damage such as a cut, burn, or fracture does not need pain to be rapidly detected by the victim. Appendicitis had been diagnosed in her by the signs of fever, inflammation, and gut motility, even though she had no pain. Unusual accidents do occur in such people in novel situations. For example, as a child in the deep Canadian winter, she climbed up to look out of the window and knelt on a hot radiator. One could still see line scars on her knees as an adult.

How then do such children learn to avoid harmful stimuli? They learn very rapidly from the alarm of others, from their teaching, and from shame. The alarm of others even affects animals. Sheep and pigs explore novel objects with their wet noses. When a thin electrical wire is stretched around a field, a few animals briefly nudge the wire, after which the entire flock keep their distance from the wire. There is no point in spanking a child with congenital analgesia to reinforce orders with pain, but they normally learn to detect anger in adults. Twins with congenital analgesia were studied in detail by specialist child psychologists who discovered that personal and social development appeared completely normal throughout their childhood. It is true that these children have to be warned not to push their luck in the more violent games that children adore. I examined a teenage woman with congenital analgesia who was a champion trampolinist. I was relieved that her enthusiasm for this sport was fading.

The Canadian student with congenital analgesia died at the age of twenty-two from osteomyelitis. Why? The history of these cases is that the analgesia begins to fade as they grow older. They begin to report deep pains, such as headaches, toothaches, and period pains, and later to feel superficial pain. For some, this spontaneous recovery comes too late. Why? We have already described the two phases of pain after twisting an ankle. The first sharp brief pain does not occur in those with congenital analgesia. However, anyone knows that they have twisted an ankle by clues that are not painful and that we use to learn to avoid the situation. Furthermore, when we stumble without pain, we are alerted, we are scared, and we learn.

What is far more important is the second phase of pain, in which the pain is deep and spreading. In this phase, far from jerking the foot away, we hold it still, guard it, and limp in order not to put pressure on it. All of us have minor accidents several times a year, often so minor that we may forget them, but, during the recovery time, we guard the damaged area, protect it, and move it as little as possible. That motor behavior, which is the opposite of the sudden brief withdrawal, is crucial for recovery because the area of damage cannot complete the inflammatory and recovery processes if it is moving and under pressure.

With congenital analgesia, this recovery phase with guarding after minor injury does not occur. The consequence is that the surfaces of joints and ligaments never fully recover. Furthermore, the joint is in particularly bad shape to counteract the next trivial injury. Strangely, a severe injury such as a fracture does not have such severe consequences because the damaged limb is put in plaster and held stationary until healing occurs. The consequence of repetitive minor injuries with congenital analgesia is that joints, particularly ankles, knees, and wrists, become demolished. The dead and damaged tissue becomes a haven in which bacteria can flourish and eat their way through the bone into the marrow. This explosive invasion, called osteomyelitis, is still extremely difficult to treat, even with antibiotics, because it is difficult for the medicine to penetrate into this hidden horror. This was the cause of death for the Canadian student and others.

We understand from the experience of children with congenital analgesia that pain has a protective role, not so much in its very acute phase associated with movement but in the prolonged phase associated with stillness. Even in the acute phase, the muscle reaction is not isolated to simply withdrawing the injured part. You cannot withdraw your foot from a sharp stone unless you stiffen the other leg. If you did not, it would be a disaster because you would collapse back onto the source of your pain. These widespread reactions become much more obvious in the second phase. Joints are splinted by the highly unusual, steady, simultaneous contractions of all the muscles that can move the joint. This contributes to the stiffness that plagues people with arthritis. It is most obvious in the ubiquitous back pains in which the long muscles on both sides of the spine are in steady contraction in their attempt to stop the vertebrae moving. You have only to look at the stiff twisted postures in a crowd of people walking in the street to diagnose those of our fellow citizens with back pain. Orthopedic surgeons make 90 percent of their diagnoses by watching their patients walk into the examination room. Dogs are a wonderful example of the widespread readjustment of muscles produced by a small injury to one foot. They switch effortlessly to a three-legged gait with one leg steadily flexed. This requires an instant reorganization of all the leg and body muscles. And so it does with us.

Overall Body Responses

The victim of sudden injury cries out. Every species has its version of this call. For some, it is an alarm call and the group scatters; for some it is a distress call for help and others rush to their aid. One-day-old baby mice emit an ultrasonic call far beyond our hearing range, and the mother runs over while the rest of the pack alert. We humans have an ambiguous reaction. "Good" citizens move to help while others deliberately move away or waver.

The entire body switches to an emergency mode. There are obvious changes in blood pressure, heart rate, sweating, and breathing. But there are also less obvious changes. A group of elephants browsing at ease on the plains of Kenya can be heard a mile away because of the loud rumbling in their gut as they digest food. Complete silence descends abruptly if they are alarmed because part of their response is that gut movements cease. Gut paralysis and constipation are examples of the widespread consequences of wounding and can even be a serious complication in postoperative patients. Elective surgery offers a particularly fruitful way to study these overall body reactions because the patient can be studied in detail before the operation and during recovery. In addition to the pain-producing nerve impulses generated in the damaged tissue, there are many other factors including anxiety, starvation, sleep disturbance, blood loss, and infection, which all disturb the body. In preparation for the emergency and the subsequent repair, the various hormone systems of the body go into high gear. They increase inflammation and mobilize sugar, fat, and protein.

A special fraction of this hormone response is surprising. Hans Kosterlitz, of Austrian origin, was one of many refugee scientists to enrich British and American science. He spent a lifetime working in Aberdeen in Scotland on the action of narcotics. He followed the dictum of Claude Bernard in nineteenth-century Paris that plant poisons led us to discover crucial mechanisms already existing in the body. Bernard's triumph started with curare, which he showed interrupted the normal passage of nerve impulses from motor nerves to muscles and led to the discovery of the natural chemical, acetylcholine, that makes muscles contract. Kosterlitz examined opium and believed it must be acting on a body mechanism that normally responded to opiumlike compounds. Finally, in his late seventies with his student John Hughes, he discovered that the brain did indeed make naturally occurring narcotics, which they named endorphins.

At almost the same time, Sol Snyder and his student Candace Pert at Johns Hopkins University in Baltimore discovered that some of the brain's cells contained highly specific receptor systems on which narcotics that changed the cells' properties attached. It was discovered to everyone's surprise that the body makes its own narcotics, endorphins, and has receptors to react to them. Among other actions, these naturally occurring substances counteract pain. A special version of these endorphins, called enkephalin, increases in the upper part of the dorsal horn among the small cells (figure 8). Furthermore, receptors for narcotics appear in damaged tissue.

This provides the basis for the most useful treatment for the pain of injury, the administration of narcotics such as morphine. Morphine simply imitates and enhances part of the ongoing process by which the body reacts to injury. Morphine combined with the natural narcotics of the body reduces pain and other reactions, such as anxiety and hormone responses. Even constipation is caused when narcotics are administered, just as it occurs when wounding releases the body's own narcotics. When nerve impulses from the wounded tissue at the site of the operation are prevented from entering the cord by local anesthetics in a spinal block, not only does the pain not occur but most of the hormone changes are also prevented.

Fainting is an extreme example of the body's overall reaction to an emergency. Nerve and hormonal mechanisms divert blood to the vital deep organs and away from the skin, which turns white. The blood pressure drops suddenly, reducing bleeding. The person is unconscious and still.

Conscious awareness of pain is one part of a massive synchronized series of reactions associated with tissue damage. The reaction is a coordinated whole. It would be very unwise to assume that each component of this overall reaction, including pain, has its own separate and private mechanism. Unfortunately, that is precisely the traditional view I will attempt to counteract.

Images in the Brain

Recent astonishing technical advances have been made that allow a three-dimensional analysis of the location of activity in the brains of conscious humans. Since 1991, a number of technically ingenious papers have been published that show which parts of the brain are active when people are in pain. The method to be discussed here is called positron emission tomography (PET). It involves the injection into a blood vessel of a solution containing a very short-lived type of radioactive oxygen. The new substance flows to the brain where its presence can be located precisely by the physical events that signal the radioactive decay of the oxygen. All parts of the brain are supplied by a steady flow of blood. When nerve cells become active, they need an increased supply of oxygen. To achieve this, the blood vessels dilate and there is an increased supply of blood to the active nerve cells. The PET scanner can detect these local increases of blood flow and register their location and intensity (figure 10).

These machines are inevitably very expensive, not least because they have to be associated with a cyclotron, which makes the very unstable short-lived radioactive oxygen. Because the method involves injecting a radioactive compound, the subject's safety requires an injection of only a small dose of atoms that must lose their radioactivity in minutes. Any method has limitations. PET scans can measure nervous activity only indirectly, by way of the blood flow. They can resolve location down to about half a cubic centimeter. The time resolution is poor, in the range of seconds to minutes, so the method cannot measure a rapid sequence of neural events. For safety reasons, the tests cannot be repeated. Yet, despite these limitations, PET scans have already produced fascinating pictures. For the future, quite new methods are being developed that are safer, cheaper, and more accurate in space and time. One of these is called functional nuclear magnetic resonance imaging (fMRI).

Every major PET center in the world has published pictures of the distribution of activity in the brain when the subject is in pain. The pictures are spectacular signs of contemporary technology, but their interpretation by puzzled brain experts has often lurched back into the nineteenth century, where each bit of the brain was given a separate functional label. They clearly expected a modern version of Descartes's picture, figure 4, where the pain center would be distinctly isolated and identified. The answer they observed was a plethora of zones of activity. The classical pathway for pain enters the forebrain at the thalamus. As expected in the classical model, painful stimuli to normal volunteers show the thalamus to be activated. However, when patients in steady pain were examined, there was less than normal resting activity in the thalamus. This seemed ridiculous to classical thinkers but is, of course, exactly what is expected if descending control circuits are attempting to limit the pain. The next station in the classical plan is the sensory cortex, but here the new results were chaotic, with some finding no change, some a decrease, and some an increase. Let us stop following classical expectations and show the result.

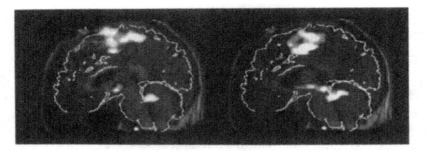

FIGURE 10. Positron emission tomography (PET) scans made by the team at the Karolinska Institute, described in the text. It shows active areas in the brain when a small painful injection was made in the arm of healthy volunteers. The image of the brain is in the same alignment as that in the top part of figure 9. From left to right, activity can be seen in the frontal lobes, in the motor and sensory areas, in the anterior cingulate, in the hypothalamus, midbrain, and cerebellum. (From Hsieh et al., *Pain*, 63, 225–236; 1995.)

For example, a group in the Karolinska Hospital in Stockholm had injected a small amount of alcohol in the upper arm of healthy volunteers. This Stockholm group is led by Martin Ingvar, who comes from a distinguished line of three generations of neurologists. (His father, David Ingvar, employing a predecessor of the PET technique that used radioactive xenon, was the first to show local increases of blood flow in the brains of people in pain). The Stockholm subjects felt pain, anxiety, and unpleasantness and their heart rate went up. The PET scan detected clear signs of increased activity in the sensory cortex and in the motor cortex. However, there were also substantial areas lit up in other cortical areas in the frontal lobes, in mid-brain areas, in some areas of buried cortex, and in a midline area called the anterior cingulate (AC in upper figure 9). These four areas had never before been thought to be directly concerned with pain. Results of this type were repeated with various painful stimuli in normal volunteers by several groups in Europe and North America.

Puzzled by these unexpected results in normal people, the research turned to patients in pain, suffering from heart pain, tooth extraction, migraine attacks, and so on. Similar widespread scattered patterns of brain activity were found in these patients. The Stockholm group carried out a particularly subtle trial on a special group of patients. When a single nerve in the arm is injured, some patients suffer a continuous burning pain in the area supplied by the nerve. Injecting a local anesthetic around the nerve produces a pain-free period. A group of such patients, who had been in steady pain for many months, were PET-scanned once and then again when the pain had been abolished by injecting their damaged nerve with local anesthetic. Surely, when the pain-free picture was subtracted from the pain-present picture, the result should show precisely those areas signaling pain. The result was approximately the same as that seen in normal people in pain.

How did the classical thinkers cope with this embarrassment of riches? They said it was too simple to expect only a single area because pain was associated with attention, orientation, blood pressure changes, misery, and so on. Therefore, they proceeded to label each area of activity with its special function: the midline area for attention, the middle area for orientation, the buried cortex for blood pressure, and the frontal lobes for misery. This is a highly satisfactory way of explaining the new data in terms of localization of function and is marred only by the fact that there is not a scrap of evidence for these fantasy labels.

There are two active areas where labeling might be plausible. One is the hypothalamus (6 in upper figure 9), an area of the brain that dominates our inner-body processes, such as temperature, blood pressure, and heart rate, and therefore could reasonably be involved in the overall reaction. However, this activity is normally thought to be a reaction to pain and therefore not itself an indicator of the pain that precedes it and triggers it. Another area thought to be involved after the onset of pain is the periaqueductal gray area in the middle of the mid-brain (4 in figure 9), which is the origin of one of the descending control pathways that reach the spinal cord, as discussed in chapter 3. There is good evidence that activity in this area turns off pain, and it is therefore a poor candidate to be labeled as the cause of continuous pain.

Quite the most revolutionary aspect of all the new data is the intense activation in structures previously labeled as having a purely motor function. These parts are the motor cortex, the basal ganglia, and the cerebellum. The defenders of classical thinking dismiss this remarkable fact by saying that movement follows pain and therefore these structures are involved in the avoidance movements. This so-called explanation cannot be true because many of the patients, particularly those who have been in steady pain for years, show no signs of movement related to their pain. The results are so surprising that we may need a fundamental shift in the target that we are seeking to locate the mechanism for pain. We naturally think in steps. First we have sensation, followed by perception with its identification, classification, and emotion, and last, perhaps, motor action and behavior. To match these three steps, classical thinking assigned separate functions to three parts of the brain: the sensory brain, the perception brain, and the motor-planning and action brain.

Now we have to face these completely paradoxical PET scans. Some patients show every sign of perception of their pain but are not moving or even planning to move, yet parts of their brain previously assigned to the motor step are intensely active. Could it be that we have made a fundamental error in expecting a sensory box separate from the motor-planning box? Could it be that we in fact sense objects in terms of what we might do about them? Could it be that we have erected an artificial frontier between a sensory brain and a motor-planning brain that does not in fact exist? We will return to these questions in chapter 10. In the meantime, we have seen here that the feeling of pain coincides with changes in every part of the body and in a distributed pattern in parts of the brain.

In this chapter we have seen that, when pain strikes as a conscious event, it is accompanied by new activity in every part of the body, including many areas in the brain. There are signs of alertness, orientation, attention, and exploration. Muscles contract to avoid the stimulus and, later, to guard the wound and aid recovery by preventing movement. The tissues of the body are altered by changes of blood flow and of hormones. It is a premature conclusion to separate conscious pain from all these other activities. It could be that the pain is the combined activity of the many groups of nerve cells. In the next chapter, we will look at the individual personal variations of this combined activity.

A "Normal" Pain Response

In chapter 2, I wrote that the most common question I am asked is whether I am working on physical or mental pain. I have given reasons for doubting the validity of that dualistic question. The second most common question is "Why is it that some people are resistant to pain?" I understand the question and why it is asked. All of us fear pain and doubt our own ability to withstand it. That is to say, we fear the onset of private pain and doubt our ability to endure it with dignity in our public display. No one approaches death or danger without this fear predominating. At the back of our minds, we all have the image of some imagined stoic who bravely copes and says, "It only hurts when I laugh."

I began to doubt the stability of this stereotype while I was still a medical student. I was assigned to take a blood sample from a massive sergeant major in the guards with the Military Medal ribbon on his tunic. Unwisely, I sat him upright in a chair and knelt in front of him to insert a needle in an arm vein. As the blood entered the syringe, he fainted and collapsed on top of me. I learned two lessons: that one should stand to one side of a vertical patient and that bravery depends on the situation. We need, therefore, to explore what is stable and what is variable in us, and what is the nature of the variation.

Most of the machinery of the body operates in a stable range in all normal people at rest. That stability is the basis of the standard clinical examination. Measurements of blood pressure, blood sugar, body temperature, and so on are standardized and are similar from Patagonia to Baffin Island, in astronauts in orbit and in submarine crews on a deep dive. Similarly, the acuity of the senses is the same among normal people wherever they are tested. Vision and hearing thresholds are similar in all peoples. A special word is assigned to the sharp feeling of a pin prick in all languages. The threshold for feeling the temperature of water as it heats up from warm to painfully hot is about the same for everyone.

A very accurate test of pain threshold to electricity is to place electrodes on the skin of a volunteer and to slowly raise the strength of pulsed shocks. At a low level, the subject reports feeling an innocuous thump that turns to a sharp pricking sensation as the strength is raised. This is the pain threshold in the sense that it is the abrupt appearance of a new type of sensation that is mildly unpleasant but, if augmented, would eventually become intolerable. Everyone who is free from disease has the same threshold for pain in this special sense. If someone has a raised threshold for this transition from dull to sharp, then they have a disorder such as a disease of their peripheral nerves as in diabetic neuropathy. So far so good. We all have the same pain threshold. The variability appears when we perceive pain above the threshold level and when we approach what we can tolerate.

The Hero

Few of us have escaped an upbringing that displays the hero for our emulation. These admirable characters, from the Spartans to James Bond and Rambo, bear pain without flinching. Our teachers tell us that only those who are weak and cowardly succumb to pain. Pain is presented as an option that is not taken up by those with the "right stuff." I know of no country without its "hall of heroes" with tales of valor that, hopefully, typify the national character.

It is true that a few cultures have also invented antiheroes for the benefit of a decadent bourgeoisie. The Good Soldier Schweik survives the Austro-Hungarian army by excessive zeal in overobeying orders. Gunner Asch oozes his way through the German army with inspired skill. These rare examples of the antihero are exceptions to the popular image of mass heroism that typifies an ideal of behavior. Unfortunately, for most of us, there remain severe doubts that we will achieve the national standard when the time comes. So we wonder what makes heroes. Here we should demand to know if anyone is a permanent hero type or whether they exhibit heroism only in certain circumstances.

Many tribal cultures have initiation rites in which awful woundings must be born without a whimper. These remind us of the state identified by Beecher as the rare occasion where wounding has a clear advantage. The child is rewarded for stoic behavior by being admitted to adult society. North American Indians who survived such an initiation have been tested as adults for pain tolerance and have been found to show no signs of retaining the stoicism they showed during the initiation. The Masai in Kenya and Tanzania maintain these injury tests on adults to choose their leaders. There is no evidence for toughness among those who do not enter the competition for leadership.

We have no reason to doubt the existence of these ceremonies or to treat them as bizarre special instances that could occur only among so-called primitive peoples. To our shame, they occur regularly in our "advanced" societies. We learn about them in the newspapers only when they go wrong. It appears commonplace in elite military units, such as parachute troops and commando units, that, when new men or officers are to be admitted to the group, they must submit to extreme endurance tests. In the minds of such men, they may retain a lifelong pride in their achievement: "Once a Marine, always a Marine." However, like the American Indians, when these men are seen in later life as patients in civilian hospitals, they are just like all the other patients. It seems that heroism and stoicism are children of the situation rather than lifelong characteristics.

The image of the stoic, Spartan, macho hero is strictly male. One might have expected a reassessment of this ideal by the new feminists. Unfortunately, we learn only that women can be heroines too, a statement that was never in doubt. In *The Wilder Shores of Love* by Lesley Blanche, which is regarded as a pioneering book of feminism, we read the astonishing stories of nineteenth-century women who took control of their own lives. Women are presented in that book with role models who have an unbroken history of tough, successful achievement of complete control. I regret that women, like men, are saddled with an unreal unachievable ideal.

A particular version of the hero is Jesus Christ. He was condemned to suffer crucifixion, a method of execution designed to be intensely painful. For two thousand years, Christians have lived with repeated versions of Christ on the cross. The pictures generate feelings of reverence in which the concept of Christ the redeemer merges with Christ who suffers. Christ is presented in the gospels as very human, with doubts as he prays in the Garden of Gethsemane for "the cup to be taken away" and regrets when he calls on the cross, "Why hast thou forsaken me?" His suffering and sacrifice are coupled with the promise of redemption. I have no doubts that religion can provide immense comfort. We should, however, generate no confusion between the suffering and the redemption.

I accept that the sacrifice of citizens, friends, relatives, and comrades in the two world wars gave me the benefits of freedom but I do not confuse that fact with their sufferings. I do not accept the words of the lord mayor of Cork in 1920 as he starved to death in protest at the English domination of Ireland: "They will win who can suffer the most." In this statement, it appears that suffering itself is a virtue and I greatly regret that some Christians have proclaimed this fallacy. The saints and martyrs went to cruel and painful deaths as followers of Christ's example. Later, self-inflicted pain came to be an accepted route to join His kingdom. Saint Marguerite Marie Alconque (1647–1690), who founded the cult of the Sacred Heart, wrote: "Nothing but pain makes my life supportable." In our day, Pope John Paul II wrote:

What we express with the word suffering seems to be particularly essential to the nature of Man. Sharing in the sufferings of Christ is, at the same time, suffering for the Kingdom of God. In the just eyes of God, before this justice, those who share in Christ's sufferings become worthy of this Kingdom. Through their sufferings they, in a sense, pay back the boundless price of our redemption. Suffering contains, as it were, an appeal to Man's moral greatness and spiritual maturity.

Here we understand that the Pope accepts suffering and glorifies it. This powerful statement has had practical consequences in Catholic countries, particularly in treating pain in terminal cancer, where some doctors have hesitated in their treatment of pain and suffering because the treatment might intrude on the patients' act of redemption.

Everyone matches their response to that of an idealized *alter ego* role model. No one achieves that paragon of virtue in a steady state throughout life, and some of us never make it.

Pain Thresholds in Normal Volunteers

When faced with an uncomfortable puzzle such as pain, where stimulus and response seem so strangely connected, we have learned over the past two centuries to mobilize science in the belief that it will reveal order in chaos. In psychophysics, the relevant branch of science, a measured controlled pain-producing stimulus is given as heat, cold, pressure, or electricity. The pain response of a normal volunteer is measured on some scale or by their physiological changes in a precisely controlled environment.

This sounds like "real" science, and it is true that the results are reproducible. There are, however, three caveats we must consider if these results are to be applicable to pains that occur in ourselves or that we observe in the clinic. First, it is obvious that the experimenter and the subject are certain that the stimulus will not produce any prolonged injury or pain. This is a necessary restriction but makes it an artificial pain outside normal experience, where pain comes together with a packet of worry and doubt. Second, the subjects are assured by the scientist that they are able to stop the pain at any instant when they decide it is not tolerable. What is being measured is pain without suffering that can be instantly abolished. We should all be so lucky if that was the type of pain we wish to understand. Finally, there is a problem that applies equally to patients and to volunteers: they are both individuals and they are members of a culture.

Despite these caveats, the measurement of pain in these circumstances has been carried out in thousands of trials. Inherent in these trials is the concept of a pure sensation of pain liberated from perceptions and meanings. Many believe such a sensation exists. I do not. I think it an artifice generated in part by our dualistic culture and confirmed by the experimental design in which the subject has agreed to play the Cartesian game. What are the results?

A disciplined, attentive, well-trained group of volunteers report on the amount of pain they feel gradually rising from barely perceived to the intolerable limit. The curve of their rising pain is consistent and repeatable within the group. The onset of pain is the same between groups. The problem starts with the upper limit. A group of Harvard students set their upper limit of tolerable stimulus far below a group in Munich. In spite of identical experiments and the same instructions, the group had collectively decided on the upper limit that must presumably have depended on the society within which they lived. These experiments have been repeated many times, and each seems to give a lawful relation between the stimulus and the response, but toleration shifts between groups. The person conducting the experiment has an unintentioned effect on the tolerance. The results differ depending on whether the person applying the stimulus is male or female, a professor, a technician, or a fellow student.

In some experiments, a deliberate attempt is made to encourage the subject to go beyond their chosen limit. "What! You ask me to stop increasing the stimulus? Most people take much more than that!" Some individuals refuse to give their permission, but most allow a higher limit. If the instructions to the subject do not mention the word *pain*, and they are asked only to estimate how big the "stimulus" is, they give quite different responses on its size. They also do this if asked to estimate the weight or brightness of a stimulus. This emphasizes a crucial difference of pain from neutral events. You may estimate weight, blueness, or warmth without any necessity to do anything about it. For such neutral stimuli, the test can be a purely intellectual game. Pain always implies an impending threat that will eventually engulf you if it is allowed to escalate. This type of science is not useless, because it can even be used to assess analgesics, but its utility is limited to the highly unusual context in which it is necessarily performed.

Gender

What could be a more popular question for scientific investigation than finding out whether women or men have a higher threshold for pain? There is therefore a vast literature on the subject, and the results are contradictory and chaotic. Nowhere in this research is the meaning of a painful stimulus considered because the subjects are treated as pain-detecting machines in the same way that one might test a light meter. Familiarity with an event changes the meaning.

A very careful Canadian study recently showed that women had a higher threshold for heat pain whereas men had a higher threshold for painful electric shocks. In our society, women have far more experience of handling hot plates with apparent impunity while men, with their enthusiastic fiddling with car engines and electrical gadgets, are usually familiar with tingling or stabbing electric shocks. Nowhere is this difference in the common experience of men and women mentioned as a factor in the difference of thresholds. The meaning of the pain has been ignored or the subject has been trained that the particular experimental pain has no meaningful consequences.

Related to tolerance and meaning is the level at which one complains in public. This is measured in practice by the stage of disease at which the doctor is visited. Western men are famous for neglecting themselves on this measure. Prostate disease is a very common occurrence in men. Enlargement of the prostate is almost universal as men age, and prostatic carcinoma is the third most common cancer. Every doctor is all too familiar with the male patient on his first visit who reports getting up several times a night to urinate, a prolonged and painful effort to urinate, and wetting his pants. The condition has been developing for years, but the man excuses his late request for help on such silly grounds that he did not want to bother the doctor or he could cope or it was not all that bad. Most Western women who develop a problem urinating with frequency and urgency are round to the doctor like a shot. For whatever the cultural reason, women seek help at an earlier stage of a disease than do men.

Of course, men and women have some diseases specific to them, but they also have a lot in common. In all Western societies, women visit doctors more frequently than men, and this is not entirely caused by the additional problems of feminine anatomy. Over a lifetime, women are healthier than men in the sense that they live longer.

The lower threshold for complaint to others has practical social consequences. In a large general hospital, female nurses shared the responsibilities for postoperative patients in both male and female wards. It was found that the consumption of analgesics was much higher in the male wards than in the female wards. The nurses were carefully observed and interviewed. Their consistent attitude was that if a male patient complained of pain it must be serious because everyone knew that male patients were a tough lot and should be taken seriously. On the other hand, they had a different attitude to their fellow females, who were generally considered by these nurses to make a great fuss about minor problems and therefore were to be brushed off with a minimal response. The poor women were victims of their own social stereotype.

Genetics

Wherever there is a variation in a species or between species, it is obvious that genetics should be investigated as a possible source. One of the puzzles of pain in humans is that severe nerve injury can produce long-lasting intractable pain of great ferocity. However, an additional puzzle is that the prolonged pain occurs only in a fraction of those with apparently similar injuries. In military amputees, all have obvious serious problems but only 10 to 20 percent have the devastating pains caused by injury to their large nerves.

At the Hebrew University in Jerusalem, a twenty-year study has been carried out on rats who begin to show possible signs of pain two weeks after their sciatic nerve was cut. The animals were not allowed to suffer and were killed as soon as they showed the first signs. It was noted that some strains of rat never suffered this condition. By selection, cross-breeding, and DNA analysis, a single gene responsible for the condition has now been located. The next stage, which is being vigorously pursued in universities and pharmaceutical companies, is to identify the responsible protein. It is clear that a very specific treatment might emerge that would control the pains caused by nerve damage.

Some strains of mice have high thresholds to the ordinary pain tests. Some painful diseases clearly have a genetic origin. Sickle-cell anemia, which is common in people of African origin, is a genetic disease that provides resistance to malaria in those with one gene but produces a disease with acutely painful crises in the minority who have two of the genes.

Although genetics plays some role in every aspect of human biology, there is no evidence that it plays a large part in common pains. The main reason for this definite statement is that pain varies in the same individual, depending on the circumstances. This variation cannot be entirely due to genes as such because it is learned during the individual's lifetime.

Cultural Stereotypes

A popular sport is to mock women or men, the old or the young, mothers-in-law and so on. Close to these as favored targets for mockery are ethnic groups. If it is your group, you admire it and mock the others. The whole world is populated by peoples assigned a place on a spectrum from extreme stoicism to utter wimpishness. Icelanders mock Danes, Swedes mock Norwegians, and I bet the Cook Islanders think the Sandwich Islanders a pretty cowardly lot and vice versa. Pain tolerance is a standard popularly assigned to ethnic groups.

There have been many studies attempting to provide support for these beliefs. My favorite was a study in Ontario that set out to compare Canadian-born citizens of European origin and of Chinese origin. They could find no difference between the two groups and decided they had studied the wrong Chinese people. They then recruited recent Chinese immigrants who had been born in China. These showed themselves to be much more sensitive to painful stimuli than the Canadian-born Chinese. The reasonable explanation is that Canadians are blasé about men in white coats sticking electrodes on them whereas the recent immigrants were reasonably anxious and ill at ease.

While ethnic differences of personal sensitivity are in doubt, there is no doubt that public display is bound by culture. This has great practical consequences. For example, in the 1930s, Grantley Dick-Read was a colonial doctor in the Tulkarm region of northern Kenya. He wrote of witnessing childbirth in local women who were quiet, calm, dignified, and conversing with their tribal neighbors. He was inspired by this and wrote the book *Childbirth Without Fear*, which was to have a revolutionary effect on prenatal training classes and the education of mothers-to-be in the West. For him, the Kenyans showed that pain did not occur in the absence of fear, anxiety, tension, and ignorance. He set about showing that a mother who was transported by herself and others into a "natural" setting benefited greatly.

So far so good. But, fifty years later, a female anthropologist who could speak Tulkarm witnessed a similar scene to that which had so changed Dick-Read and the nature of prenatal training. She asked the woman after the delivery whether it had hurt. The woman answered that the pain had been great. She was then asked why she had not said so and replied: "That is not the custom of my people." There really should be no surprise at this answer. The women in labor rooms in Oslo do not shout out not because they are not in pain but because it is not the custom of their people. At the other end of Europe, in Naples, women in labor do shout very loudly not because they are in more pain than the Norwegians but because those around them would be worried if they did not shout.

Needless to say, people are consciously aware of the stereotype of their group and, when in doubt, may act accordingly. The Irish may put on an Irish act, begorrah. The English may put on their stiff upper lip, don't you know. I was sent to find an English soldier who had been admitted in the middle of the night to the emergency ward of a French hospital in Lyons with a broken leg after being run over by a truck. I had no difficulty in locating him because he was shouting, crying, and cursing in English, surrounded by French-speaking staff who were attempting to calm him. I walked up to him and said, "What's up, mate?" The effect was as though I had given him an anesthetic. He switched in an instant to the cockney "It's nuffink, doc. I'll be all right." With three words, I had transported him from the isolation of an alien culture to the familiar.

In 1969, Mark Zborowski wrote *People in Pain*, comparing pain reactions in American residents who were white Anglo-Saxon Protestants (WASPs) with those of Irish, Jewish, or Italian origin. He claims that the intensity of their reactions varies in that order. It now reads as the opposite of politically correct and is the stuff of novelists, giving a characteristic label to people of differing ethnic groups, although Zborowski's book is written in "psycho-sociology speak."

At about the same time, the psychologist Robert Sternbach and colleagues carried out a more controlled study on volunteer women from the same four groups. They found somewhat similar group differences but went further. When challenged in a pain tolerance test to accept a larger stimulus, some Jews would permit a considerable escalation of the acceptable stimulus whereas most of the women of Italian origin refused to budge. More important, they noticed that some of the women were adopting quite different tactics. The old Americans tended to "roll with the punches" while the others made themselves rigid in anticipation of each stimulus.

This variation of tactics in response to an identical stimulus can be seen dramatically in a line of young soldiers who are about to receive a vaccination shot. Some sweat and tremble. Some appear calm. Some distract themselves by chattering with their mates. Some look at the needle while others look away. Some faint. We witness here the diversity that characterizes the individual at that moment. This expression of personality influenced by culture and experience is not unique to humans. The phrase "led like a lamb to the slaughter" implies an innocent uninformed ignorance and passivity for impending horror. However, the fact is that sheep standing in the pen outside the slaughterhouse differ from sheep in a field. They are quiet and they do not eat or drink the available food and water. Most do not react to being touched. Some explore and probe. Some, "sheeplike," trot along behind the first into the slaughterhouse while others hang back and panic and have to be wrestled inside.

Cultural stereotypes have a limited validity and show that humans and animals have a wide variety of options and tactics that are particularly apparent in painful situations. In an intensive study of cross-rearing, it has been shown that a terrier who has been suckled by a spaniel mother yelps and rolls over when bitten by any normally reared terrier. Yet the same dog dominates his adopted spaniel brothers and sisters. The stereotype is a myth of generalization and is not even stable. The antisemitic labeling of the Jew as "coward" can hardly survive the performance of the Israeli army.

Hypnosis

Normal people under hypnosis may declare that their hand is comfortably warm when dipped in ice cold water. Ever since its origins with Franz Mesmer in the eighteenth century, hypnosis has been the subject of an avalanche of mystical blather. The French Academy set up a commission to investigate it, with Benjamin Franklin, who was an ambassador in Paris at the time, as a member. Since this was the "age of reason," the commission declared that mesmerism and animal magnetism did not exist because they were beyond reason. The commission was wrong. The confusion over hypnosis stems from a belief that hypnosis necessarily involves a person in a trance state induced with some hokum maneuvers by someone with special powers. None of these are necessary. The essence of the hypnotic state is only that the subject has agreed to hand over to the hypnotist the responsibility for deciding how they will react and what they will sense.

This is a common condition of everyday life. It is called the acceptance of authority. The lawyer cross-examining a witness in court and massaging the testimony until the witness makes the required answer is a hypnotist. The parent telling the bedtime story of how the family dog is a retired ballet dancer is a hypnotist. In the traditional hypnotic setting, a form of contract is agreed between subject and hypnotist. Since this is voluntarily accepted, there is no conflict of the subject's standards and therefore the possibility of the subject acting evilly, as in the Svengali myth, is out of the question. Because the process depends on an agreed transfer of authority, the attitude of the subject will crucially depend on their personal, cultural, social, hierarchical relation to the hypnotist: "Does this person impress?"

True anarchists are terrible subjects. If the subject faces an innocent suggestion, such as "Your left arm is beginning to feel heavier and heavier, so heavy that when I tell you to try, you can't lift it," almost everyone responds. As the task becomes more improbable, and perhaps threatening, fewer respond. Even in good subjects, long practice and experience are necessary, particularly where pain is the trial.

Hypnosis was widely used in the first half of the nineteenth century for surgery before the days of general anesthesia and as a substitute for rum, laudanum, and physical restraint. The patient was prepared in long-repeated sessions with the hypnotist. Even so, there were some horrific scenes when the patient broke through the pain-free state. A famous physician was dismissed from University College Hospital in London in 1840 when a woman broke out of hypnosis during a mastectomy. She sued both hospital and doctor. When the first general anesthetic was administered in London, the surgeon, Joseph Lister, stepped back at the end of the operation and declared: "Gentlemen, the Yankee trick beats the French one." He was referring to the import of ether anesthesia from Boston being more effective than the hypnosis associated with Mesmer in Paris.

Acupuncture during surgery has all the characteristics of hypnosis. I was invited to China in the mid 1970s to witness this fascinating phenomenon because the Chinese thought that the "gate control" theory of pain, which Melzack and I had invented, was a Western version of their theory. With three other doctors, I followed some twenty patients from the patients' ward through the operation and back. The patients, who had volunteered for acupuncture rather than general anesthesia, had all been through a long course of training. They were at ease with the acupuncturist in a trusting brotherly way. They had experienced many trials and, because they shared the ward with other patients, they were familiar with the whole surgical procedure. On arrival in the operating room, the atmosphere was relaxed and friendly with everyone greeting everyone else.

Some of the operations were difficult to assess because, in addition to needles, the patients received large intravenous doses of narcotics and often local anesthesia in difficult areas. However, two patients calmly went through major surgical incisions without any additional medication. The question of a hypnotic explanation for the whole phenomenon arose sharply in one patient who had the femoral artery explored and cannulated *before* the needling started. The patient was calm and conversing. She knew she had been assured and convinced that she was not going to be hurt even before the acupuncture.

The similarity to hypnosis increased in a much more unpleasant episode. The patient was being operated on for the removal of one lobe of his lung, which involved a wide opening of his chest while acupuncture was proceeding. The operation went well with the patient being given oxygen and some sedatives. The operation was approaching its end and the incision was being closed. As almost the last step, a drain was pushed from inside his chest through the chest wall. The man screamed and struggled to get off the table. He was held down and emitted a babble of screaming, crying, and shouting. This was exactly the type of breakthrough described in the last century. I believe they had talked him through each stage of the operation and had followed the plan they had previously taught him. The insertion of the drain was not in the plan and caught him unawares, and he reacted with extreme pain.

I wish to make it absolutely clear at this point that the relief of pain by acupuncture is a real phenomenon. That should not be in question, even though it is true that it was used in its pure form on only a small minority of patients in the Chinese hospitals we visited. The majority of patients were operated on with standard Western-type general or local anesthesia. Even for the minority who had acupuncture needles inserted, the effect was usually augmented by substantial doses of intravenous narcotics and by local anesthesia of sensitive points. The question about acupuncture is not about whether it is sometimes effective but rather about the mechanism of the action. The classical Chinese opinion that the effect results from the needles affecting a biological energy flowing in special channels has no foundation. I propose here that acupuncture anesthesia has some similarities with hypnotic anesthesia. In chapter 9, we will return to examine the subject in detail, describing the placebo, which can be highly effective and depends on the subject's learned expectations.

Hypnosis was used for two centuries in an air of mysticism and bafflement. In recent years, sober studies have begun to explore its real nature with three startling results. When a normal person places their hand in ice cold water, it hurts and, as we have described, there are a whole series of reactions other than the verbal report, such as changes in the heart rate and increased blood pressure. If a well-trained, experienced, responsive subject under hypnosis is told that the water is warm, they agree that it is warm even though it is, in fact, ice cold. What happens to their blood pressure and heart rate? They react exactly as if the water were ice cold. In other words, the person is divided into separate parts. The speaking person says that the water is warm, but the body's nervous system, which handles internal regulation, detects ice cold water and reacts correctly. This division places hypnosis on a specialized verbal plane, since we have already shown that the body normally acts as an integrated whole.

The next stage of discovery involves a further division. It starts with the work of the Hilgards at Stamford, one a professor of psychology and the other of pediatrics. Their very well-trained, responsive subjects were hypnotized and instructed so that each subject becomes two: a speaking subject and a writing subject. The speaking person was instructed that the water is warm even though it is iced, and obediently declares it warm. The writing subject, also called the "hidden witness," had not been told the temperature and was asked to write what is felt, and writes that the water is ice cold. It is apparent that a very sophisticated game is being played in the relation between the hypnotist and the obliging subject.

This reminds us of people with multiple personalities, suffering from the Morton-Prince syndrome, who have also been shown to multiply their personalities at the suggestion of an authority figure. The discovery of this authoritarian aspect of hypnosis has led to its decreasing use by psychotherapists, who do not relish simply ordering their patients to change. Obeying orders may look very impressive at first, but obedience tends to fade with time. There are an awful lot of cigarette smokers who gave up their habit after hypnosis but drifted back to smoking. The same is unfortunately true for patients treated for pain by hypnotic suggestion.

The third group of studies in relation to pain and hypnosis has been done by neurologists in Paris. They are experts in evoking pain in volunteers under strict control by electrically stimulating a nerve in the leg and precisely measuring the reflex withdrawal of the leg. In general, the stronger the shock, the stronger the pain and the stronger the reflex. They then take well-trained responsive hypnotic subjects and suggest that the pain will decrease, and the subjects accordingly report that the pain is less. They have made no suggestion about the reflex. Half the subjects who report a decrease in pain also produce a decreased reflex. However, the other half who also reported an equal decrease of pain produce a markedly increased reflex. I take this crucial experiment to amplify the variability of tactics by which we respond to painful stimuli. Even here, where one fraction of the overall response was being manipulated, the other fractions were adjusting with differing strategies. I have not set out to show that pain is a single entity with a simple bottom line. When pain strikes, the individual has many options that are intended to end the pain.

Images

For all the variation of response discussed in this chapter, it might seem reasonable to believe that somewhere in the head there is a big flashing light that says *PAIN* and that this would be in the same location in everyone in pain. This is an allegory, but in actual recorded data one might expect a certain group of nerve cells to become more active and to be the coded signal associated with our feelings and doings that we call pain as a shorthand expression for the associated activity. In the last chapter, we looked at examples of activity in various parts of the brain associated with pain states. They were obtained by the PET scanning method, which produces fascinating results but has weaknesses. To protect the subject from radiation, it is necessary to inject only small amounts and so the detected signal is weak. To raise the signals out of the noise, it has been necessary to add the results from many individuals. Obviously, this has not permitted an answer to the question of individual differences.

I mentioned a new method, fMRI, which does not involve radioactivity, has a better time and spatial resolution, and permits the analysis of the amounts of neural activity in individuals. The first results are just appearing, and they are a shock. Twelve normal individuals were scanned while they were receiving brief, hot, painful stimuli to one finger. As we have come to expect, the amount of pain reported by the subjects varied widely (from 1.5 to 8 out of 10), despite the fact that the stimuli were identical. If all twelve individuals are pooled, they show increased activity in the same group of structures that had been shown to be active in the PET studies. However, if the individuals are examined one by one, no two individuals have the same pattern of activity. Of course, this is a serious shock to the traditional thinkers who expected specific, specialized brain structures that could be named as the location of the flashing sign.

The result is not as hopeless as traditional thinkers might feel. They may be a sign of each individual adopting a personal tactic for dealing with the stimulus. We have looked in this chapter at the variety of options that are apparent in a group of individuals. In the same individual with repeated identical stimuli, the tactics may change. Everyone has had that experience at the dentist. I was a guinea pig for an antimalarial drug trial that involved my giving daily blood samples for weeks. I grew to hate the needles and used every trick I could invent to cope with the brief pain of the intravenous needle. The traditional expectation is that neural codes exist in the rate and amount of activity of groups of cells. The more the cells fire, the bigger will be the signal. Of course, there are many other possible codes, and there is evidence for some of them. For example, there might be a time code where first one group fires and then another. The time resolution of the contemporary imaging methods is not yet quite good enough to achieve such a search.

There is another more subtle version of the time code in which assemblies of cells could group together in such a way that only synchronous impulses count as letters in the code. Much of modern computer software is concerned with codes, and each variation suggests ways in which the brain might operate. A nightmare for the neuroscientist would be the existence of shifting codes because these are hard to crack. We should not be depressed that the most advanced modern techniques fail to show a single simple focus of brain activity associated with pain. Pain may be described as a single simple word, but it implies a class of responses involving many areas of our brains and bodies. The pattern of response varies from person to person, and within an individual it varies from one painful episode to another.

Pain After Surgery

I place this in a chapter on normal pain in normal people because it is so common when we observe in ourselves and in patients and friends that the reactions change from one person to another, even after apparently identical operations. Preparation for an operation inevitably involves rising tension and anxiety. Entry into the hospital involves a rite of passage to translate the person from free citizen to dependent patient. Forms are filled in with an implicit threat. Next of kin and religion are requested. A permission form is signed that transfers responsibility to others. The patient is stripped of familiar clothes and dressed in a silly gown in a strange room with strange people. Even the life-giving, familiar morning coffee is forbidden. In "good" hospitals, an attempt is made to ease the patient's growing puzzlement by explaining step by step what will happen. This rarely lasts for more than a few minutes and contrasts with the long course of easy familiarization that I described in Chinese hospitals, where the patients are recruited to be members of their own treatment team. The onset of general anesthesia is a blessed relief, not only from the pain during the operation but from the hurly-burly of the preparation, with its fatigue, fear, anxiety, depression, and loss of sleep, all of which carry over into the postoperative period and influence pain.

After the operation, most patients are in pain, but there is a huge variation. There is no such creature as a standard patient, even after identical operations by the same surgical teams. Fortunately, there have been great advances in recent years such that patients can expect and even demand comfort. The former inadequacies of postoperative treatment came from two sources. One was the failure to treat the patient as an individual but to give medication in predetermined doses at set intervals. The other was to give medicine in minimal doses at the longest intervals because doctors and nurses misunderstood the dangers of overdosing and were afraid of creating addiction. A more realistic acceptance of the variability in people has led to more honest monitoring and more open belief in what the patient says. In search of safety, patients in the past were given short-acting drugs, and, after a fixed interval, they were permitted a second dose when the first was no longer acting. This produced particular misery as the patient cycled between comfort and pain with no helpful response by the staff to their painful phases until the time came for the next dose. Now the ideal is achievable by prolonged or continuous analgesia and by steady monitoring of the patient.

One of the particular miseries of postoperative patients is their helplessness, a condition rare in normal life. Infancy reappears and adds shame. If the recurrence of pain means that the patient must attract the attention of a nurse, who has to ask a doctor for permission to give analgesia, there is an all-round increase of tension and irritation. For exactly that reason, patient-controlled analgesia was invented in which the patient can administer their own additional doses through a machine with built-in safety controls. The effect is to restore control to the formerly helpless patient and, unsurprisingly, the total amount of analgesic medicine that brings pain relief to the patient is usually less than the amount that would have been given by the staff.

There is no time in a hurried hospital routine to diagnose why some patients are in more pain than others. It is true that some operations, opening the chest for example, tend to be more painful than others, such as opening the abdomen. However, the causes of the variations in pain go far beyond these simple mechanical reasons. Every sensible surgical patient has good reason to be fearful, anxious, and depressed. The intensity of these feelings will affect the intensity of the pain. No amount of psychotherapy or drug therapy will abolish these entirely reasonable emotions, but they help. A particularly difficult feature to measure is the patient's personal assessment of the meaning of the operation in terms of what the future will bring. A patient may not believe the euphoric assurances of the most charismatic surgeon who declares that everything is in order.

Beyond the period of pain, there are often surprisingly long periods of fatigue, depression, and malaise. Some of these are detectable before the operation. They too do not have simple explanations attributable to the disease, the anesthetic, or the surgery. In a recent study, it was shown that recovery from hip replacement is particularly brisk despite the fact that this is a very major operation involving considerable tissue damage. It is suspected that patients have an extremely optimistic attitude to the outcome of this operation with a good chance of pain relief for their arthritis and for greatly improved walking. In other types of operation, the patient may have a more pessimistic view of the future, which they express in prolonged depression and malaise.

This chapter has been about the variability of people. It is deep in human nature that we respond individually to any threat, including pain. Our internal variation is compounded by the attitude of others who impose their stereotypical cultural expectations. There are external sources that influence the amount of pain. These include authority and the cultural stereotypes of the hero, ethnic groups, and gender. In addition, the amount of pain is influenced by internal states that are themselves often affected by external events. These include fear, familiarity, expectation, depression, and anxiety. Behind the personal variable reaction lies the origin of the pain itself, to which the next two chapters are devoted.

Pain with Obvious Causes

In this chapter, the immediate causes of pain and the way in which they develop will be described. There will be no reference to the variation in the amount of pain observed in real life, as discussed in the last chapter. In later chapters, we will bring together the basic cause of the variation.

We will discuss a selection of common pains, covering the spectrum, from a scratch to an amputation.

A Scratch

A scratch, by definition, is a trivial injury, but it contains many of the pain-producing components found in much more serious conditions. At the instant of the scratch, pressure has opened channels in the nerve fibers in the skin and generates nerve impulses that begin their travel to the spinal cord. Pressure, cold, heat, and chemicals are the four ways in which sensory nerve impulses can be initiated. The scratch produces an immediate stinging sensation due to pressure, but this is only the first stage of a process that now proceeds (and was described in chapter 3). Inspection of the scratch at the time reveals a white line. Within seconds, the line becomes red and the redness spreads slowly in a flare. This is not bleeding into the skin but an opening up of the small vessels in the damaged skin and the nearby area. Even later, a bump appears in the skin along the line of the scratch. This is the weal and is caused by clear fluid and cells moving from the blood vessels into the damaged tissue. Here, on a tiny scale, are the four signs of inflammation: redness, heat, swelling, and pain.

The pain changes from the initial sting to an ache and the area, if gently pressed, is tender. Tenderness arises from two causes. Chemicals leaking from the damaged tissue have sensitized nerve fibers. This means that chemical stimulation has been added to the temporary initial pressure stimulus and prolongs the pain. The second cause of the tenderness is that cells in the spinal cord have been woken up by the initial volley of nerve impulses and are on guard. Two stages may follow the scratch with its initial phases of flare, weal, pain, and tenderness. If bacteria have been introduced, local infection grossly exaggerates the four signs of inflammation: the color becomes angry, the swelling weeps clear fluid, which may become milky with white cells, and the pain grows. If infection does not occur, the repair process eats up the cell debris and cells migrate in that stitch the edges of the skin together to produce the white line as the mark of an old scratch. The pain goes.

A Twisted Ankle

To recapitulate chapter 3 in brief, a twisted ankle has all the characteristics just described for a scratch but is bigger and plays out its course in three dimensions. The initial sharp pain is followed by a deep spreading pain. The ankle becomes hot and swollen. Here are the cardinal signs of inflammation with which we are now familiar. However, there is a new phase in addition to the local signs. The whole foot and lower leg are tender. The leg hurts if moved and is guarded and not permitted to touch the ground or bear weight.

The pain now originates not only from the original site of injury but also from tissue distant from the injury. A touch to undamaged tissue that would normally be felt as innocuous pressure now feels painful. The origin of this distant pain comes from the readjustment of spinal cord circuits to produce a widespread area of pain well beyond the small local injury. The effect is to prevent any movement of the injured part and thereby to facilitate recovery. The pain that located the site of injury in the first phase now spreads widely and is associated with immobilization of the body part.

A Toothache

A toothache starts with inflammation entirely inside the tooth and is signaled to the brain by the nerve in the root of the tooth. Peculiarly, the pain may seem to come from a distant part, for example an inflamed molar back tooth may produce an earache. This is called referred pain and is an error of localization that we will discuss again in the section on heart attacks. It is different from the spread of pain just described in the twisted ankle.

If the inflammation spreads out of the tooth and involves the gums, the pain changes its quality and is precisely localized to where the problem is. It becomes obvious as an area of red, swollen, and painful gum and face. A third pain may then appear with a quite different origin. The tissues of the body are drained by a massive network of little channels that converge into the lymph nodes. The lymph nodes that receive the drainage from the mouth are in the neck under the jaw. As the debris and bacteria from the inflamed tooth and gums are filtered by the lymph nodes, they too may become inflamed, swollen, and painful. Finally, the breakdown products of the damaged cells may leak into the bloodstream and produce a fever with malaise and general aches. Local damage can evidently produce distant pain by these three mechanisms of referred pain, lymph nodes, and fever.

A Heart Attack

Each heartbeat is produced by a vigorous wave of contraction of heart muscle that sweeps over the entire heart. This contraction requires energy, provided by a steady supply of blood delivered by the coronary arteries. A heart attack occurs when one of these arteries is suddenly blocked by a blood clot. The precise sequence of events has now been followed in detail when patients are under close observation in an intensive care unit. Blocking a coronary artery immediately deprives a wedge of heart muscle of its blood supply. It is known from studies in animals that this produces a vigorous discharge in the fine nerve fibers that supply the heart and deliver the messages to the spinal cord. The intensity of the volley of nerve impulses rises in seconds and then falls off in minutes.

The immediate reaction of patients is surprisingly diffuse. They report a deep stomach discomfort but also a feeling that something is very, very wrong. I have spoken to two friends who are pain doctors who have had the modern operation in which a balloon is pushed into one coronary artery to widen it. They both report no pain to their surprise but an awful feeling of intense terror of impending death. The fine nerve fibers from the heart do not connect to nerve cells that permit precise localization but they do set off a general feeling of anguish, nausea, sweating, and breathlessness.

This is just the beginning for the heart attack patient because the arriving volley of nerve impulses in the spinal cord sets up a slowly spreading excitation of nearby nerve cells that normally react to the chest, arms, and neck. As this "bush fire" spreads, the patient reports something large slowly expanding in the chest, reaching the surface of the chest like a giant clamp, and spreading down one arm, usually the left, and up into the neck with agonizing pain. Of course, at this stage the progress of the pain may be brought under control by the use of analgesics.

Here we have looked at the natural history of a very specific pain, angina pectoris, in which no pain is located in the heart itself, all the pain being referred to structures whose reporting nerve cells lie close to the cells receiving from the heart's nerve fibers. Angina may occur repeatedly in less dramatic circumstances in which the coronary arteries are partially clogged. Here the blood supply to the heart is adequate at rest but, when the patient begins to exercise, the heart has to beat more frequently and more strongly. The narrowed arteries cannot supply enough blood for the increased demand, the nerve fibers in the heart signal that the oxygen supply is inadequate, and the patient feels the typical pains of angina until he rests or takes medicine to boost the heart's blood supply.

Osteoarthritis

Almost everyone in their seventies is creaky in the joints. The beautiful smooth joint surfaces of their youth have simply been worn away by continuous use. The more heavily used joints, which have been under most pressure, such as hips, knees, and the joints between the vertebrae, have ragged rough surfaces that are very obvious in X-ray pictures. In an attempt to repair and stabilize the joints, cells move in and make tough fibrous tissue that stiffens the joints. Damaged cells produce the familiar inflammatory response, which spreads to surrounding tissue, making it tender. The pain may be referred to distant areas, so osteoarthritis of the hip may be reported as pain in the upper leg and knee. As was described for the twisted ankle, a chronic defensive tactic to encourage recovery is to immobilize the joint. Eventually, unused muscles wither.

As with the other conditions described, the natural history of local tissue damage spreads its effect, including pain, to nearby and distant structures by a linked series of separate mechanisms. This is nowhere more clear than in osteoarthritis, in which the disorder starts locally in a joint but spreads to nearby tissue while the nervous system, attempting to immobilize the joint, generates an abnormal gait and posture, which in turn stresses other joints.

Childbirth

This is an example of pain without illness. During pregnancy, the growing weight of the baby plus the uterus, placenta, and amniotic fluid eventually add up to about 16 pounds (8 kilograms). This inevitably produces a change in posture, often with backache and sometimes nerve compression pain in the legs. For many months the baby has been floating free in the amniotic fluid and is felt to be kicking and sometimes somersaulting after about sixteen weeks.

As the time of birth approaches for a normal delivery, the baby's head descends into the lower part of the uterus, a cone leading to the cervix, and becomes fixed. In abnormal births, the body and bottom sometimes fit into this space, and there is trouble ahead with a breech delivery as the body and legs are delivered first. At this time, the uterus begins to build up quite strong contractions that the mother can feel, but without pain. The growing pressure bursts the membranes and the amniotic fluid drains out. The contractions now grow in frequency and intensity. The contractions of the uterus press directly on the baby's body and force its head into the cervix. The cervix is normally closed, but the head forces it open until it is about 4 inches (10 cm) wide. The opening of the cervix is the first cause of pain, which the mother feels with each contraction, the pain centered in the lower abdomen, often extending in a band around the whole body from front to back and sometimes radiating down the legs. Between contractions there is a steady pain in the abdomen and back.

After this first stage of labor, the baby's head passes through the cervix into the vagina and begins to advance more rapidly. It presses on the vagina, bladder, and rectum and then bulges out the floor of the pelvis, called the perineum. Since the location of the pressure has changed, the location of the pain shifts down to the pelvis and perineum. In order to speed up this particularly painful stage, forceps may be put on the baby's head and a cut, called an episiotomy, may be made to enlarge the opening of the vagina. After the delivery of the baby, the intensity of the pain drops dramatically. However, depending on the amount of inevitable tissue damage produced by the passage of the baby, prolonged inflammatory processes may now start up, producing long periods of soreness.

At the end of chapter 3, we looked at the McGill pain questionnaire. The sensory words most commonly used by mothers to describe the pain during labor are *sharp, cramping, aching, throbbing, stabbing, hot, shooting, tight,* and *heavy.* For the emotional affective feelings, the most common words were *tiring* and *exhausting.* A rating scale for pain intensity was used, with a number assigned by each mother. The average rating for women having a first baby was 35, and for those who had previous children the number was 30. These numbers are on a scale in which people with broken bones rate their pain at 20 and cancer patients at 27. Scores above 35 were reported only in cases of nerve injury or amputation. Evidently, the average reported pain of childbirth ranks high in the range of human experience. Of course, there is an extremely wide variation in the reported intensity and I have discussed variation elsewhere. But it is crucial to remember that, of first births in Canada, 9.2 percent of mothers described their pains as "mild," 29.5 percent as "moderate," 37.9 percent as "severe," and 23.4 percent as "excruciating." A study of Scandinavian women (despite their reputation for toughness and stoicism) yielded similar results.

The cause of pain in the first stage of labor is the generation of nerve impulses in fine pressure-sensitive sensory fibers in the uterus and particularly the cervix. These impulses excite the spinal cord in the lower back, and all of the pain is referred to distant structures. Just as was described for the heart, the pain is not located in the uterus or cervix. As the baby's head advances into the vagina and presses on the pelvic floor, nerve impulses arrive in the sacral spinal cord and activate cells to produce a pain that is correctly located in the structures being squeezed.

Cancer

There are only two good things to be said about cancer: some can be cured and some are not painful. Cancer is completely different from the painful conditions we have just described in which the cause of the pain is associated with tissue that has been smashed and is in the very abnormal state that causes inflammation. Cancer cells are almost exactly like normal cells except that they are multiplying. For this reason, the body does not recognize them as foreigners and therefore sets up none of the normal defensive inflammatory responses to react to them. This means that cancer in its early stages is painless except on very rare occasions. This is a terrible disadvantage because the lack of pain signals at this initial phase means that diagnosis is delayed.

The body has an elaborate and precise mechanism for recognizing foreign objects, including cells. This defense mechanism explains why transplants, or grafting of tissue from other people, never works unless the immune system is suppressed. Cancer cells grow by stealth, accepted by neighbors as normal. They may grow as a solid lump or may invade between neighboring cells and infiltrate over long distances. Cancer cells may float free in the bloodstream or in the ducts of the lymphatic system and lodge elsewhere to establish colonies called metastases. During this stage, the victims are unaware that they are host to a silent invader.

In a recent case, a high-ranking air force officer made a faulty landing on a carrier deck. As the plane careered out of control, he activated his ejector seat and was safely rescued. Because firing the ejector seat rockets puts very heavy g-forces on the body, it is routine to carry out an X-ray examination of the entire body to search for small bone fractures that may have occurred. In this routine scan, it was discovered that this apparently healthy man, who had recently passed a rigorous medical check, had a sizeable brain tumor that was producing no signs and, certainly, no pain.

If cancer is silent in its early stages, why does it have such a reputation as a painful disorder? The answer is that the swelling lump eventually causes pressure. As brain tumors grow they block the drainage of the fluid in which the brain floats, pressure rises in the head, and headaches result. Tumors in the gut grow to such a size that they block the normal passage and painful cramps result. Tumors may block blood vessels, producing pain in tissue starved of oxygen. Lung cancer becomes painful only when it blocks small air passages and local inflammation attempts to remove the sick area of lung. Bone tumors are painful because they press on the sensitive covering of bone. They may also produce pain by weakening the bone so that it fractures.

These secondary pressure effects explain at least 80 percent of cancer pains. Surgery may produce a dramatic beneficial effect on these pains by relieving the pressure block even if it does not cure the cancer. Similarly, X-rays and chemotherapy may have excellent palliative effects by shrinking the lumps. Only a few quite rare cancers, such as those produced by asbestos, emit pain-producing substances, making the cancers painful by themselves before they reach the pressure-pain stage.

One type of cancer pain comes from nerve damage that itself is caused by pressure when the cancer infiltrates major nerves. We will discuss why this is so painful in the next section on amputation, which causes the major problems from nerve damage. Unfortunately, pain may be a side effect of treatment. This is a crucial area of the relationship between doctors and patients in which a full explanation must be given about the balance between the chance that therapy will cure or ameliorate the cancer and the chance that therapy will produce devastating side effects. Radical surgery followed by radiation treatment to remove an entire breast with its cancer and all the lymph tissue in the armpit, which may contain spreading cancer, gives the patients a chance that they may survive five or ten years or longer. However, while patients decide to take this gamble, they must understand there is a chance that the treatment may also leave them with disabling pain. We carry out this grim balancing of risk versus benefit throughout our lives, but there is no time when an honest, open assessment is needed more than when a patient considers treatment for his or her cancer.

Cancer pain is worse than useless. It provides absolutely no protective signal because the disease is far advanced before it starts. Once started, it announces the obvious and, if it goes untreated, it simply adds to the miseries of impending death. Worse, untreated pain accelerates death. Fortunately, the great majority of these pains can now be treated to bring real comfort to the dying patient.

Amputation

In this chapter, it has become apparent that the causes of pain migrate as the condition develops. These sequential changes are particularly evident when a limb is amputated. The common picture of amputation is of a violent event in warfare or in accidents. However, amputation becomes more common in peaceful and aging societies where it is necessary to remove a leg by surgery because the circulation has failed owing to disease of the blood vessels. The immediate pain of amputation is the consequence of the extensive tissue damage. We have described that sequence repeatedly in this chapter with the step-by-step reaction and recovery from damaged tissue. However, a novel painful element appears because amputation necessarily involves cutting the major nerves that supplied the limb.

An immediate and bizarre consequence is the appearance of the phantom limb as a clear insistent sensation that convinces the patient the limb is still present in every detail. This phantom limb is an invention of the brain, which has lost the normal steady sensory input from the limb. When the input fails, the cells that receive the input raise their excitability in an attempt to seek the missing input. The consequence is that the brain receives false signals as though the limb was still present. The sensation at the early stage is not painful, but it rivets attention because the brain senses that there is something highly unusual happening.

As time goes by, the nerve fibers that have been cut attempt to grow back into the missing limb. The cut ends send out delicate sprouts that end up in a tangle because they have nowhere to go. These new fibers begin to generate spontaneous nerve impulses even though there is no stimulus. The patient feels tingling in the phantom. The young sprouts are unstable and respond easily to pressure. This produces painful tender spots on the stump. Because the sprouts try to grow out into the missing limb for the rest of the patient's life, these tender areas remain.

When the nerve was cut in the amputation, a massive injury barrage swept into the spinal cord. This generates the spreading tenderness even in undamaged tissue, just as we described for the twisted ankle. However, in addition, the outgrowing nerve sprouts find themselves in foreign territory quite different chemically from the areas in which they previously ended. The sprouts pick up these unusual chemicals and transport them to the spinal cord. The cells there get a second signal from these chemicals that something is seriously wrong, even after they have responded to the first signal by way of nerve impulses. The cells set about reorganizing in an attempt to react appropriately. They become more and more excitable, sometimes firing steadily, sometimes firing off in great coordinated explosions, and often overreacting to sensory signals arriving from the remaining uninjured part of the limb.

The patient feels the consequences of these cells attempting to restore normalcy. Some feel a steady ache in the phantom that can rise to a horrific sense of burning and a feeling that the missing foot or hand is in a permanent cramp. Some feel violent stabs of pain. The nature and intensity of these sensations vary from patient to patient. In everyone, gentle touching of the stump and remaining limb reveals patches of skin that are exquisitely sensitive and painful, with the sensation often radiating into the phantom. Unlike the painful signs of inflammation, which die down as the wound heals, these awful sensations may continue with little modification for the rest of an amputee's life.

Although ongoing and episodic pain with zones of painful tenderness are obvious in many amputees, they can occur in any disease in which nerves are damaged. Shingles (herpes zoster) is a condition that becomes more and more common with age. The acute stage is purely inflammatory. Chickenpox viruses, which have been hibernating in one of the dorsal root ganglia since the patient had chickenpox as a child, suddenly multiply in vast numbers. The viruses migrate along the sensory nerves to the skin supplied by the ganglion and produce a band of redness with swelling, pain, and often a fever. All these signs of inflammation, with which we are now familiar, die down leaving scarred skin. However, some of the sensory nerves have been destroyed, and some patients develop all of the miseries suffered by amputees but limited to the area where the nerves are damaged. The skin aches and is exquisitely tender and there is a miserable deep ache. If this second phase follows the acute inflammation, the patient is likely to be in trouble for the rest of her or his life. Therapy is poor.

What Other People Say

In this chapter I have deliberately described pain in a standard way without dwelling on the huge variations. It is, of course, a complete myth that a standard cause produces a standard pain. However, the myth affects the reactions of patients, medical staff, and friends alike.

If you walk into an accident and emergency department with a broken wrist, you are matching the pain you feel with the pain you expect to feel with a broken wrist. They rarely match. You may be astonished if it does not hurt. You feel additionally miserable if your pain reduces you to a helpless weeping wretch. The nurses and doctors instantly spot that you have a wrist at a strange angle, make a diagnosis, and assign you to a category. For them, a broken wrist comes with an appropriate package of pain and deserves a fixed protocol of action. You as an individual may be anxious, terrified, drunk, or sad, but for them you become "the broken wrist in cubicle 6." When your friends turn up, they too add to this soup of disparate expected, "appropriate" behaviors. They may act relieved that it is "only" a broken wrist. They may complicate your sadness by weeping with you. Their sympathy and care help or hinder your own progression of feelings, assessment, and worries. It is all a genuine drama with each actor seeking his or her appropriate role in the play entitled *The Man with the Broken Wrist*. Some people are wonderful in such a situation, some are paralyzed, and others hopeless. Everyone influences the patient's pain.

Emergency situations are a problem, but the real challenge occurs when pain persists beyond the "appropriate" time. Medical personnel can become grossly dismissive and switch their attitude to suspicion that this is no longer a "proper" patient. They may act as though pain is an option, especially when the initial pathology has gone but the pain continues. The doctor becomes uneasy when patients fail to respond to their treatment and this discomfort of the doctors may escalate into irritation, guilt, anger, and withdrawal. Friends who are not doctors play their role with sympathy, support, shelter, and care, but these too wear thin if the patient's pain fails to diminish along the expected path. As this situation advances, all players including the patient begin to share guilt, anxiety, and depression with major scenes of pity and self-pity, defiance, and resignation. A very different reaction is needed by all involved in this chronic phase than during the early reactions to the beginning of pain.

In this chapter we have seen that the location of most pains caused by obvious damage coincide with the area of damage but then extend as a result of the spread of inflammation and the raised excitability of spinal cord cells. When the damage occurs in deep organs, such as the heart or uterus, the pain is not located in the damaged organ but is referred to a distant zone on the surface of the body. When nerve fibers are cut across, there is a particular danger that the spinal cord cells on which these fibers terminate react to the loss of input by becoming hyperexcitable. Cancer pains are a special case because the body does not recognize cancer cells as being foreign and so the cancer increases without pain and inflammation until the cancer has grown to such a size that it produces mechanical damage and the consequent inflammation or nerve damage.

In all of these instances, the nature of the pain is influenced by the patient's attitude and that of her or his friends and doctors. If the pain persists, the attitudes of patients, doctors, and friends are tried progressively more severely. That challenge is even worse where there is no obvious cause for the pain, which is the subject of the next chapter.

7

Pain Without a Cause

For the past two hundred years, the very considerable advances of academic medicine can be attributed to the insistence on identifying a clearly defined cause for each disease. Before the modern era, causes were often mystical and there was little attempt to verify them. The proposals that the patient was suffering from imbalanced humors had been accepted for two thousand years. Treatments that worked were justified by attributing their action to adjustment of the same mystical forces.

Our language retains the words from the old medicine invented by Galen, who was born in the second century A.D. in Pergamum in Asia Minor. He believed that all disease was caused by an imbalance of the four elements (earth, air, fire, and water) with the four cardinal humors of the body (blood, phlegm, bile, and black bile). This scheme dominated medicine for the next eighteen hundred years, and we still speak of rheumatism (too much water), pyrexia or fever (too much fire), pneumonia (not enough air), cholera (too much bile), and melancholia (too much black bile). Galen was not unique in this way of thinking as the Chinese had a scheme based on the imbalance of yin and yang, and astrological doctors could explain all disease by the conflicting influence of the planets.

By the eighteenth century, mystical causes were largely dismissed in favor of causes defined in the new scientific terms. Pathology and physiology began to locate disordered tissue and to explain how the function of the body could enter disordered states. The discoveries of bacteria, viruses, and chemical errors identified ultimate causes for the local abnormalities of tissue. The entire canon of modern medicine became pathology driven. A raft of diseases are now understandable as the consequence of exactly defined pathology and therefore rational therapy emerges.

Unfortunately, a large number of common painful conditions do not have an associated pathology and so represent a severe challenge to both doctors and patients. Many doctors are so impressed with the power of modern pathology that they refuse to accept the existence of disease without pathology. This attitude hugely exaggerates the problems of patients who suffer from pains that are not considered "real." Some doctors take the hopeful approach that, while the pathology has not yet been discovered, future work will reveal the basic cause. Even Sigmund Freud wrote that neurosis would eventually turn out to be a biochemical disorder but, in the meantime, he proposed that psychoanalysis was a productive approach. Other types of doctors, deeply frustrated by their inability to cope with diseases without pathology, turn on the patient and claim that the pains are self-inflicted by a faulty way of thinking. I will now examine a series of painful conditions that have no known cause and will describe both the conditions and their possible causes.

Trigeminal Neuralgia

We will start with a fairly rare condition that causes severe pain so specific that one would think that there must be a local pathology, and yet none has been discovered. Tic douloureux, as trigeminal neuralgia is also known, is a terrible pain in the face with stabbing pain on one side often described as a strong electric shock. Between the attacks, the patient is pain free and normal. The pain comes on abruptly and stops equally abruptly. The pain explodes if a gentle moving stimulus brushes a trigger point, usually in the area around the mouth. It occurs equally in men and women and is most common between the ages of fifty to seventy years but may occur at any age. If nothing is done, the pain may disappear for months or years only to reappear in the same area. The little trigger point that sets off the stab of pain may meander slowly from one part of the face to another.

It would seem entirely reasonable to patient and doctor that there must be some damage to the nerve fibers leading from the trigger point to the brain. However, this is not so reasonable when we consider the following facts. The nerves in the face have been examined in great detail and no abnormality has been found. The sensitivity of the skin of the trigger point is completely normal between attacks, and, yet, if nerves are damaged, one would expect abnormal sensation. Finally, as we have said, the trigger point meanders about while nerves remain in the same place for a lifetime.

The nerves from the face originate from cells in a ganglion in the base of the skull that is the equivalent of the dorsal root ganglia that lie on either side of the vertebrae and give rise to sensory nerves to the body. From the ganglion, the nerve fibers group into a root, which makes its way into the brainstem where the fibers connect with central nerve cells. The root runs free in the fluid around the brain. One theory proposes that the nerves are damaged by being pounded with pulses from nearby blood vessels. There may be some sign of minor damage, but just such damage is present in many people who have no signs of the disease. In the brainstem, there are no signs of damage in most patients, but some patients, who have multiple sclerosis in the brainstem area where the face nerve fibers arrive, do develop tic.

Fortunately for patients, two methods of treatment are used, and they tell us a lot about the nature of the disorder. The French word *tic* means a "twitch" and is also used for an epileptic attack. With no more logic than the similar use of the word for both conditions, antiepileptic drugs were given to patients with trigeminal neuralgia and found to be highly effective in the majority of patients. This is good for the patients but shifts the search for the cause of the disease. These drugs have no known effect on peripheral nerves but certainly affect abnormal firing of central cells, raising the possibility that the cause lies in the brain. Another quite different drug called tocainide also has no known peripheral action but affects the brain and stops trigeminal neuralgia. Unfortunately, some patients cannot tolerate these drugs because they cause nausea, dizziness, somnolence, and confusion, and these patients are offered surgery.

There is no doubt that the trigger for the pain originates in the large, low-threshold nerve fibers in a small area of skin. These can be temporarily blocked by local anesthetic, which interrupts the disease. For a longer-term effect, surgeons cut the nerves going to the trigger area, which stops the pain until nerves regenerate and the pain returns. In order to stop the regeneration, surgeons move centrally to attack the ganglia that contain the cell bodies of nerve fibers. Destruction here prevents regeneration because the cell bodies from which the nerve fibers grow have been killed. In the process of treating the ganglia, it was found that surprisingly gentle manipulations could also stop the pain on occasions.

We are left with two possible causes of the disease. One could be that there is some undiscovered and unknown process in the peripheral nerve that grossly amplifies the incoming signal. The other is that the brainstem cells that receive the sensory fibers have developed an extreme hyperexcitability so they explode in synchronous action when stimulated by a normally innocuous input. The beneficial effect of surgery for this central cause would come from blocking the input. The beneficial effect of the drugs would come from decreasing the excitability of the central cells. We know that cells can become hyperexcitable because that occurs in epilepsy, and we have described how pain-producing cells can become hyperexcitable after inflammation or nerve damage.

Trigeminal neuralgia is a disease without a known cause, but the target area within which the cause should be sought is obvious. The cause might be in disordered peripheral nerves or in central cells.

Headache

You may reasonably be annoyed that I place headache in a chapter on pains without a cause if you know that your own headaches are caused by irritation, anxiety, fatigue, alcohol, and so on. You are quite right, but we need to understand why those conditions set off some mechanism in the head that makes a head ache.

There are many types of headache without any structural lesion to explain the cause. Migraine and tension headaches are very common, while others are rare but devastating, such as cluster headaches, or even very rare such as ice cream headache or headache triggered by sexual activity. Migraine occurs in about 15 percent of any population studied. Environment can hardly be considered a deep cause because rates are found to be the same in country and city dwellers, graduates of Oxford and long-term prisoners. Rates are higher in women, especially after puberty, but men have identical types of attack. Before an attack, there may be hours or days of vague premonitory symptoms such as mood changes, yawning, or craving for special foods. This may be followed by an aura, most commonly visual with shimmering lights and patterns, which then proceeds to a headache on one side (the word migraine comes from the French *demi-craine*, which means "half the head"). This pain grows, often with pounding and tenderness accompanied by a dislike of light and sounds, and leads to vomiting and sleep. Once the attack is over, patients are relatively normal although subtle changes remain that could be due to medicines taken or to the fear of the next attack.

These headaches are called vascular headaches because of the throbbing that times with the heartbeat. It is obvious that such a complex suite of signs and symptoms can hardly all have a single cause. There is no evidence that the early phase of the aura has a vascular origin, but there are many changes in the brain. If there are vascular changes, they appear late in the attack and may well be secondary to central changes. In spite of this, the rationale for the treatment of migraine has focused on the blood vessels. Drugs directed at the blood vessels include the older remedies derived from ergot and the more recent drug sumatriptan. The success of these compounds in arresting a migraine attack should not be taken as proof that the pain originates from the blood vessels. They have central actions and the vascular effects may be secondary to these.

Tension headache is the type of headache with which we are all too familiar. It can last from half an hour to seven days. It has a pressing, tightening, nonpulsating quality on both sides and does not usually stop everyday activity. It is clearly different from most migraine attacks. It has been called a tension headache because it feels as though the muscles of the head are tense. Unfortunately for this belief, there is not a scrap of evidence that the muscles are overactive.

So there is no evidence that either blood vessels in spasm or muscles in cramp cause headaches. Doctors are therefore once again driven away from the peripheral tissues where the pain seems to exist and need to search the brain for the cause. We do not yet have the ability to study brain mechanisms in sufficient detail to pin down a central cause, if it exists.

Backache

An attack of pain in the lower back has caused 60 percent of the population at some time to take more than a week off work. The number of those who suffer has increased so strikingly that the condition has become a serious economic and social problem to add to the misery of the individual victim. For most attacks, the pain dies down in one to three weeks, although others take months, and some who suffer never fully recover from the first episode. Because the numbers of people having an initial attack are so great (back pain is almost universal), the fraction who never fully recover make up very large numbers of our chronically ill. Everyone, acute or chronic, is convinced that there is damaged tissue in their back and can put a finger on the area that seems to be the origin of their problem. There are tender areas around the apparent center and the pain often radiates down the leg.

It is commonly believed that the cause of lower back pain is a slipped disc, which is believed to extrude from between the vertebrae and to press on the root carrying the sensory fibers. A slipped disc can be seen in X-rays and is present in 1 to 3 percent of the population. Slipped discs are seen with the same frequency in people in pain as in those who are not. If people in pain with a slipped disc are treated without surgery, the extrusion of the disc may or may not disappear, but this bears no relation to whether or not they are still in pain. The confusion by surgeons over the role of the disc is shown by the large variation between countries of the rate of operations to remove the extruded disc. Ten years ago, the rate per 100,000 was 100 in Great Britain, 200 in Sweden, 350 in Finland, and 900 in the United States. These rates are now dropping and mark the end of a disgraceful period in which a myth was peddled to the profit of a few and the disadvantage of many, some of whom were clearly worse off as a result of surgery.

There are five generally accepted causes of back pain. They are a slipped disc or other types of vertebral disorder, an area of infection, a tumor, a fracture, and arthritis. When patients with back pain are carefully examined, a maximum of 10 to 15 percent of them may be found to have one of these five causes. This leaves 85 percent with no apparent cause, which produces a very large social, medical, and personal problem. There are a number of other causes proposed by practitioners of alternative medicine, including misplaced vertebrae, trapped nerves, and disordered joints, but so far there has been no convincing demonstration of these causes. It would seem natural to add injury as a cause, but in the vast majority of lower back pain victims there is no evidence for injury. In large surveys of companies such as the aircraft manufacturer Boeing it has been repeatedly shown that the rate of back pain complaints is the same among clerical workers as among factory workers who lift heavy weights. There is therefore no evidence that heavy or unusual exercise leads to lower back pain.

It is not true that the nonspecific lower back pain patients "only say" they are in pain. Their posture is abnormal and some muscles are in steady contraction, which tilts the back into an unusual shape. Movement is not free and there is palpable stiffness. These muscle contractions could be secondary to the pain if the muscles are attempting to splint the back to prevent pain-producing movement. These prolonged contractions could spread the pain so it migrates from its early location to settle in new structures.

Fortunately for most patients with sudden onset back pain, the condition dies down within three to six weeks with minimal treatment. Patients should be permitted a day or two in bed followed by graded activity that speeds recovery, even though it is painful. They are also given minor analgesics, usually of the aspirin family. Depending on the society, they may also receive a variety of types of physiotherapy, acupuncture, yoga, manipulation, osteopathy, and chiropractic. These latter therapies have been studied intensively and, although they may produce impressive temporary effects, they have not yet been shown to have a long-lasting effect on recovery.

The abandoned patient whose pain continues but in whom no damaged tissue can be detected is in serious trouble. The doctor may say, "There is nothing wrong with you. It is all in your head." The patient is forced to consider that he is the cause of his own suffering and is completely puzzled. If the word spreads to friends, relatives, union organizers, employers, and social security offices, his loneliness is extreme. If the doctor's message spreads to others as "Don't encourage him. It will only make it worse," the patient exists in a near vacuum. In any Western country, there are considerable numbers of people who identify the major problem of their society as being a huge mass of swindlers and manipulators who are deliberately stealing money from social services to live in luxury and do no work. The patient is now assigned to this pariah status.

People in pain have difficulty coping. Pain monopolizes their world. Anger, fear, rejection, and isolation clearly make matters worse. The patients' conflicting sense of shame and dependency adds to their problems. When these have been deliberately triggered by the authority of doctors, the patients are in deep trouble. Very rarely, individuals can diagnose the social situation, stop complaining, and put on an act that signals to others that they are not in pain despite their continued problems. These rare individuals may appear admirable, but their public performance is a sham. For the rest of us, we need comfort, support, recognition, and help if we are to make the best of our days in pain. To achieve that effort, we certainly do not need a group of doctors to wash their hands of us and dump the problem entirely on our shoulders.

Repetitive Stress Injury

This is another example of a painful condition that is epidemic in its increase and yet appears to be without cause. It occurs particularly in office workers who use typewriter keyboards or computer mice. It has not suddenly appeared but used to be given names such as *writers' cramp* and *tennis elbow*. It appears first as a tingling or numbness in the fingers. As it develops, it becomes so painful that the hand cannot be used and long rests are needed before it dies down. The pain may spread up the arm and invade the shoulder and neck. Pain and tenderness may persist after very long periods of rest. The victim becomes unable to work. The standard medical examination reveals no cause.

There is a very similar disease called carpal tunnel syndrome that has almost identical symptoms. However, in this case a test to measure the ability of nerves to conduct from the hand to the arm shows that they are conducting slowly and nerve entrapment is suspected. The two major nerves to the hand could become trapped where they pass under a band of fibrous tissue on the front of the wrist. This band of tissue holds the nerves, tendons, and blood vessels close to the bones of the wrist and is called the carpal tunnel. It is believed that, with heavy continuous use, the band swell and straps the nerves. A simple operation cuts the band and liberates the nerves from the pressure on them, sometimes completely curing the condition.

Repetitive stress injury (RSI) not only has no obvious signs of damage but seems to occur in outbreaks in particular workplaces. This tended to make industrial doctors, employers, and insurance companies suspicious that they were dealing with some form of mass hysteria of the type we have described for lower back pain. Attempts to solve the problem were made by calling in ergonomic specialists who adjusted the tilt of the keyboards, put backs on the chairs, tilted the seats, and changed the lighting. Although temporary effects were achieved, the workforce returned to their old problems. In two large telephone companies in the United States, one had serious problems but the other had few complaints, despite the fact that the equipment and tasks were identical. This observation exaggerated the belief that the disease was in the mind rather than the body. The belief was further supported when a difference was noted between the two companies. The one with problems paid workers on commission for each call answered, while in the other the workers were on a weekly wage.

The growing belief that it was "all in the head" received a further boost when a large publishing company recognized that, unlike the so-called work-shy, movement-intolerant workers of Boeing complaining of back pain, the workers who complained of RSI tended to be eager-beaver ambitious types, while their more happy-go-lucky, slobbish workmates had fewer complaints. All of this had led to angry confrontations in many countries between workers with the condition and compensation agencies who refused to recognize this as a "real" disease because they believed it had no organic cause.

Very recently, workers with RSI have been examined in London using an unusual but simple neurological test. The palm was tested for the ability to detect a carefully controlled vibration. To the great surprise of doctors and scientists, the ability of the RSI patients to detect vibration was greatly reduced on the palm of the hand but was completely normal on the back of the hand. Furthermore, when they tested workers from the same company with the same job who were not complaining of RSI, they found that many of them were beginning to lose their ability to detect vibration. These tests immediately put an entirely new interpretation on the disorder because they provide evidence for a subtle disorder of the peripheral nerves in the hands and arms or in the brain cells that detect vibration. If some of these nerve fibers or cells are not functioning properly, the disorder is reclassified as one of nerve damage rather than as a psychosocial disorder. It is well known that in some cases of overt single-nerve damage the pain spreads far beyond the territory of the damaged nerve. This means that the wide spread of pain on the body of these RSI patients joins a well-known category of pain, rather than being a bizarre distribution that had previously made unfriendly doctors suspicious that RSI was simply a form of hysteria.

We are left with a question. If RSI is now to be awarded the honor of having a "real" cause rather than a self-inflicted psychosocial cause, why has it become more common? It could be that, in the past, complainers were simply dismissed, particularly as the majority were women, but women have now learned to fight. The numbers of people who regularly type as part of their job has greatly increased with the appearance of computers and with the near disappearance of the specialist typist. And there has been a subtle change with computer keyboards. The carriage return of the old-fashioned typewriter, which gave the typist a brief rest at the end of every line, has disappeared. The use of the mouse forces the worker into very fast, detailed movements and puts them almost into the category of violinists. Musicians have been famously plagued with painful cramps similar to RSI. Finally, forced pace and the additional task of making running corrections may have increased the mechanical strain of muscles and nerves.

There are important consequences of the transition of RSI to a disease with a cause other than the psychological inadequacy of the victim.

Fibromyalgia and Myofascial Pain Syndromes

Fibromyalgia patients have widespread pain in many parts of their body, particularly in parts of the spine. They have many tender points where firm pressing with a finger produces pain. These points are particularly likely to occur in the lower back, neck, shoulders, hips, hands, knees, chest wall, and feet. In addition to feeling pain, the patients tend to be fatigued, wake unrefreshed from sleep, and be very stiff in the morning. Some 25 percent of these patients have less common complaints such as irritable bowel syndrome, headaches, and distress. The general state of fibromyalgia can be a common but temporary state.

An early morning visit to one of the airports in London reveals the sad sight of hundreds of passengers off the red eye specials from the East Coast of the United States. They are tired, lacking sleep, their biological clocks are set five hours behind, and they are irritable, anxious, and stiff. They drag along looking like a mass migration of fibromyalgia patients. Most of the travelers will recover, but for patients it is a chronic state. It is ten times more common in women than men. Muscle and tissue taken from the tender points is vaguely abnormal, but it is possible that these abnormalities result from inactivity and so are secondary to the pain. There is a suspicion that there is something wrong with their immune systems, and the condition sometimes occurs with frank disorders of the immune system such as rheumatoid arthritis.

The response to treatment is disappointing. Some patients improve when given antidepressant drugs, which are known to have an analgesic action as well as an effect on depression. A vigorous fitness training regime improves some patients, as does specific muscle training. A three-week course of cognitive behavior therapy improves the ability of some patients to cope with the miserable state.

Myofascial Pain

Unlike fibromyalgia, myofascial pain is localized to one area. There are tender points where pressing will often result in pain, which spreads to distant areas and imitates the pain of which the patient complains. Under the tender points, a taut band of muscle can be felt. The pain may be alleviated by stretching the muscle in the band, by injecting local anesthetic into it, or by pushing a needle into it. In the 1930s, some early pain experts imitated this condition in themselves by injecting a small volume of concentrated salt solution into a muscle or ligament. They felt the pain spread to a distant area from the injection site and persist for up to a day. The patients may be unable to move the muscle in which the trigger point and taut band lie or, if they do move the muscle, it provokes the pain. Myofascial trigger points may appear in the region of injury to the vertebrae as in whiplash accidents. Many studies have examined biopsies taken from the area of a trigger point and have found no abnormality. The taut band is produced by contracting muscle, but this is not strong enough to produce a cramp. In some patients, the pain lasts for up to two months and disappears with no aftereffects. There have been no controlled studies, but recovery may be speeded by treatment of the local tender spots and by exercise. When the pain has lasted for six months or more, the prognosis becomes progressively worse. Local treatment of the tender spots provides temporary relief, but the tenderness returns.

In these conditions, patients are convinced that the problem and cause must reside at the tender spot. Because no convincing abnormality has been found at the tender spots, the cycle with which the reader is now familiar starts up. Many doctors believe that it is impossible to have a local pain without a local cause. Therefore, because they can demonstrate no local cause, they conclude that the disease does not exist. This is a dogmatic throwing out of the baby with the bath water. There are logical hypotheses that could be tested to give these conditions the honor of an open investigation. For example, the taut bands could be caused by the firing of a few motor neurons in the spinal cord in a hyperexcitable area that also produced sensation. As it is, this disease with no cause is not even diagnosed in Britain with a proper title, even though doctors admit to having observed it.

Orphan Pains

There are many pains whose cause is not known. If a diligent search has been made in the periphery and no cause is found, we have seen that clinicians act as though there was only one alternative. They blame faulty thinking, which, for many classical-thinking doctors, is the same as saying that there is no cause and even no disease. They ignore a century's work on disorders of the spinal cord and brainstem and target the mind. The mind for them is foreign territory owned by the patient. Because they believe that the mind is governed by free will, they are saying that the patient has invented the pain for some devious purpose. A mountain of hogwash has been written about patients inventing pain for some symbolic purpose or as a method of manipulating other people. These are the doctors who repeat again and again to a Second World War amputee in pain that there is nothing wrong with him and that it is all in his head.

Of all the hundreds of patients I have seen, there was only one I believed had a "symbolic" pain, by which I mean that the patient uses the word *pain* to represent some other urgent need. Bill Noordenbos, a neurosurgeon in Amsterdam and the kindest, gentlest doctor I have known, introduced me to a Dutch woman with one arm amputated. She complained of a burning pain in her phantom hand but she went on to say that only one finger was burning. Neither of us had ever heard a patient describe so localized a pain so we went on to explore the details.

She had been driving her car on a hot summer day with her arm out of the window. A car driving from the opposite direction scraped along the side of her car and sliced off her arm. That evening, her husband came to see her in hospital and told her that he had no use for a one-armed wife and was leaving her. On further questioning of this poor woman, she said that her pain came from her wedding ring, which was burning her finger. She was given intensive counseling so that she came to see the loss of her arm as a tragedy but the loss of her husband as long overdue. The pain in her hand went away.

A dramatic example of an incorrect search for the location of the cause of pain is seen in patients with a damaged spinal cord who report pain in the numb part of their bodies.

Some patients who have their spinal cord cut across develop a deep, severe pain located in the completely anesthetized part of their body. There are some twenty cases in the literature in which surgeons have believed that some pain-producing nerve impulses must be traveling through their useless spinal cord and that the pain-producing nerves originate in the damaged spinal cord. They have therefore opened up the spinal cord above the area of damage and have removed completely some segments of the remaining cord. The operation has no effect on the pain and should never be done. This is a truly central pain. Some cells that have lost their normal input have reacted in an attempt to recreate their former role by raising their excitability and beginning to fire steadily. No one knows where these cells are because we do not yet have the diagnostic tools to locate them. Vigorous research is in progress to find these cells and, perhaps even more important, to understand the chemistry of their excitability, explaining what makes them run wild.

I have mocked the search for the cause of pain that jumps from a peripheral cause straight to cognitive processes, as though the only function for the massive, complex, intervening nervous system was simply a mechanical relay. Of course, that is not to say cognitive processes play no role in pain. All of us have sources of aches and pains that fail to capture our attention when we are busy and happy and yet can dominate us when we are down, lonely, and miserable. The last chapters of this book will integrate sources of pain into the vast repertoire of possible reactions that are the nature of us and our brains. This chapter has attacked simplistic solutions that propose that, if an appropriate cause for pain cannot be detected in peripheral tissue, then the pain must be an invention of the mind. There are alternative solutions to the problem of pain without a detectable peripheral cause. Diagnostic methods are not yet good enough to detect all disorders, particularly in nerves and soft tissue. Increased sensitivity of nerve cells in spinal cord and brainstem can produce "false" signals but present diagnostic methods do not yet permit the detection of such abnormal activity.

How Treatments Work

Holistic medicine proposes that pain originates from messages generated by smashed cells or disordered nerve cells that feed into the individual's brain where the messages are interpreted in the context of the person's overall needs. Classical academic medicine concentrates only on the incoming messages. In this chapter we shall look at therapies for pain whose rationale is based on stopping the incoming messages.

Testing Therapy for Effectiveness

The aim of these tests is to reveal the consistent action of the therapy liberated from the bias of patients and therapists who think it ought to work. The major bias is the placebo response, which is surprisingly powerful.

The standard test for efficiency is the randomized double-blind placebo-controlled trial. "Randomized" means that patients with a particular condition have been picked at random in order to avoid choosing some special types who might tilt the result. "Double-blind" means that neither the patient nor the people running the trial know if the real therapy or a mock therapy has been given. "Placebo-controlled" means that the true therapy and a mock imitation of the true therapy have been given.

To give an example, suppose that a pharmaceutical company wishes to market a new analgesic and that they have completed all the preliminary trials for safety and apparent beneficial effect. They are then required by law to submit their new tablet to this rigorous type of trial. A group of patients with some definite problems, a wisdom tooth extraction, for example, are asked to volunteer for a trial. They are told they will either receive the new tablet or a blank one that looks exactly the same. Then the patient, who does not know which tablet he received, tells an observer, who is also unaware of the nature of the tablet, whether the tablet reduced his pain. Finally, after all the data have been collected, the code is broken and it is calculated whether the new drug is superior to the placebo.

This type of trial sounds simple, even if it is elaborate and costly, but there are severe problems. The first problem is that the tested group never includes all the types of people who might use the therapy: a test group in New York is unlikely to contain many Inuits. The group is usually deliberately restricted, to healthy young adult men, for example. This means that the results do not necessarily apply to Inuits, women, children, old people, and so on.

Much more serious is the famous difficulty in keeping a secret. A tested drug may have side effects such as drowsiness, so it is obvious to patients and observers who has received the active therapy. There are ways around this particular problem by deliberately giving a sedative that is not believed to be an analgesic to compare with the tested analgesic that has sedative side effects. It will be seen that the crucial element of keeping patients and staff ignorant becomes more and more difficult as the therapy becomes more elaborate. Take the problem of subjecting acupuncture to a rigorous trial. What would be the placebo arm of the trial? The problems escalate to an extreme with surgical therapy, where it would be quite unethical to subject a patient to a general anesthetic and a mock operation in order to test the true efficiency of the surgery.

There are subtle ways around these problems, which we will discuss with particular therapies. However, no one should forget that the background for these trials is based on the powerful assumption that all people are the same and that individual psychosocial factors are irrelevant. This leads to the present vogue for "evidence-based medicine," driven partly by the tradition of academic medicine and partly

by the financiers' need to identify proven therapy whose cost is justified by trial. The trials are designed to identify a class of pains, medicines, and patients who respond reliably and to exclude a class where
personal individuality is a factor. The separation of these two classes is
itself an artifact because they interact. However, we will proceed as
though this generally accepted diversion permits the description of
therapies used against pain, while leaving discussion of the placebo
response to chapter 9.

Medicines

Anti-Inflammatory Medicines

This story begins in modern times with a letter from the Reverend
Edward Stone of Chipping Norton to the Royal Society in 1763. He
wrote: "There is a bark of an English tree which I have found by experience to be a powerful astringent and very efficaceous in curing aguish
and intermitting disorders." "Aguish" means rheumatism and "intermitting disorders" refers to bouts of fever. The tree was the white willow, *Salix alba*. Stone's rationale for this herbal remedy was that he and
his contemporaries believed rheumatism and bouts of fever came from
swamps and that bountiful nature would provide a remedy from the
same source. Willows grow on wet land. He mentions a similar but
different remedy, in that case for malaria, coming from "Peruvian
bark" (quinine) as another example of "the general maxim that natural maladies carry their cures along with them." This is fantasy, but the
cure works. Stone did not know that Hippocrates, Galen, and Pliny
were all advocates of willow bark. In the nineteenth century, chemists
made an extract called salicin, and MacLagan's *Lancet* paper of 1876 is
called *Treatment of Rheumatism by Salicin*.

A vast expansion in the use of salicin followed the synthesis in 1899
by the German Bayer Company of the pure compound acetylsalicylic
acid, which they named aspirin. The outbreak of the First World War
in 1914 isolated Britain from the source of this wonder drug, and the
government offered a prize of £20,000 to anyone who could find an
alternative synthesis. The prize was won by the Australian George
Nicholas, who sold his tablets as aspro. Finally, in 1971, J. R. Vane in

London showed that aspirin worked by blocking one of the pathways by which damaged cells make chemicals called prostaglandins as a crucial part of the inflammatory process.

I have told this story at some length because it shows that a crude herbal mixture was used for two thousand years, a purified extract for a century, and the precisely synthesized chemical for another seventy years before the rationale for the use of aspirin was discovered by Vane. It is intellectually satisfying to understand exactly how a remedy works, and that knowledge may lead to the development of more effective drugs. However, it is equally clear that effectiveness can be established without any idea of how or why the therapy works, or even with the wrong theory to explain its action.

Aspirin, for all its excellent properties, has its problems. Because it interrupts part of the inflammatory process, it also blocks some of the beneficial parts of that process. It decreases blood clotting and is even used for that reason as a long-term low-dose preventive treatment to reduce the chances of clotting in a heart attack or a stroke. It can interrupt the normal repair process that is going on in the lining of the stomach and, in the extreme, can produce massive bleeding in the stomach.

In an attempt to overcome these dangers, every pharmaceutical company in the world has been hard at work to produce substitutes. They have produced a vast family of sons and grandsons of aspirin. They vary in potency, but they all have the same problems. Distant relatives were added, such as paracetamol and di-pyrone. Because vast profits were to be made by renaming members of the same family, the public relations experts pushed aside the pharmacologists and spent giant advertising budgets persuading us to buy their brand. Television advertisements proclaim their cure as specific for aches and pains, but there is hardly any difference between them except in price. However, clever pharmacologists are at work to find a way to block the inflammatory pathway affected by aspirin in such a way that the bleeding does not occur. These tablets, called cyclo-oxygenase inhibitors, cox-2, are not yet available but are on their way.

Aspirin has a subtle effect on only one part of the inflammatory pathway, but it reduces pain and swelling and fever. A much more vigorous approach is to use steroids, which cancel the signals that set off the whole inflammatory process. They are used in emergency situations to bring inflammation under control, but their widespread side effects are also powerful. For example, inflammation is a crucial tool in our battle against bacteria by walling them off and destroying them. Steroids block that action and the bacteria can have a field day.

I have described inflammation as a sequence of actions and reactions with many components. Modern science is concentrating on this process, and there are hopes that it could be controlled in a beneficial way to eliminate unwanted aspects such as pain while leaving the protective, restorative functions intact. Steven McMahon in London has recently shown that a small protein that is required in the embryo for the growth of sensory nerve fibers, and is therefore called nerve growth factor, is also necessary to produce pain but not to trigger the other components of inflammation. There are therefore hopes for genuine analgesics that affect pain from the periphery but leave other parts of inflammation intact. They do not yet exist as safe, available, tested medicines. That does not mean that existing anti-inflammatory analgesics should not be used. Some people refuse to take drugs for fear of addiction, dependency, side effects, or fear that their effectiveness will fade. They are usually wrong and need help to overcome their fear.

Medicines Acting on the Central Nervous System

Opium is an herbal remedy with an even longer history than willow bark. Like aspirin, it has spawned hundreds of descendants, so aspirin and opium together are responsible for at least 95 percent of the analgesic medicines used today. Unfortunately, unlike aspirin, opium has gathered a disgraceful public image since Victorian times. For three thousand years, opium was used to produce sleep and dreams, the origin of the phrase *pipe dreams*. Although opium and its derivatives, morphine and heroin, have become associated in the puritanical mind with misuse by social dropouts, opium was used for centuries as a means of relaxation. Robert Clive, who conquered and organized India for the British, adopted the local custom and used opium regularly for the rest of his life. Where the harried New York businessman may drink his evening Martini to relax, his equally powerful rival in Singapore smokes his evening opium pipe for the same purpose.

By the nineteenth century, physicians had begun to realize that opium was not just a way of putting patients into a nearly unconscious state. By this century, it was finally realized that low doses had a purely analgesic action while leaving the patient thinking clearly. The advance can be attributed to Dame Cicely Saunders, who invented the hospice movement for the care of terminally ill patients. She and her colleague Robert Twycross decided to combine the best of modern medicine with age-old tender loving care to bring comfort to cancer patients who were dominated by their pains in the anxious shambles of their last weeks.

Doctors up to that time had joined the generally held opinion that narcotics were dangerous and that comfort was brought at the price of addiction to rapidly escalating doses, which eventually killed the patient. A cool, calm analysis of the effect of narcotics demolished this view and showed that doses carefully titrated to bring down pain to a bearable level led instead to a comfortable, clear-thinking patient. The success of carefully controlled and monitored narcotics for the benefit of cancer patients in pain spread to other problems, such as the control of postoperative pain and pain in childbirth.

The variable herbal mixture of opium was analyzed into its constituent components in the nineteenth century. The most powerful fraction was found to be morphine, which was synthesized, while a weaker component, codeine, was also found to be effective against less severe pains. This discovery set off the expected search for related compounds in the hope that one could be found that was a pure analgesic, which could not be misused as an addictive social toy and which would not stop the patient breathing when high doses were given. The 150-year search for this pure analgesic has failed, but that has not stopped the drug companies from trying. If you enjoy black humor, you may be interested in the late nineteenth-century discovery by the Bayer Company of a morphine derivative that they named *heroine as a particularly powerful narcotic they claimed was free of an addictive potential. How wrong can you be!*

Work in the past century has generated hundreds of compounds with slightly varying properties but hugely varying potency. One, called etorphine, is ten thousand times more potent than morphine. This is the drug used in darts by wildlife experts to immobilize big game. I have made it a personal rule never to remember drug doses because it was always safest to look up the dose. However, I make an exception for etorphine, for which the dose for elephants is one milligram per ton. A large elephant gently lies down if shot with a syringe containing 3.5 milligrams of etorphine. Once the veterinarians have done their job, the same dose of an antagonist is injected and within minutes the elephant stands up and wanders off with a puzzled expression.

Not surprisingly, a drug effective as a narcotic always has unwanted effects even if it is used as an analgesic. Narcotics produce constipation, and opium has been used for centuries to control diarrhea by taking advantage of this side effect of gut paralysis. High doses of narcotics depress respiration, so weak narcotics such as codeine are included in most cough mixtures. When pain fails to respond to one of the aspirinlike drugs, it is common to move the treatment to a mixture combining an aspirinlike component with a weak narcotic. The widely advertised strong over-the-counter painkillers available contain these mixtures. Much commercial and scientific ingenuity has gone into inventing mixtures that will optimize the desired effect and reduce the unwanted ones.

Only in the past twenty years has the rationale for the use of narcotics against pain become apparent. This was revealed in a series of very surprising steps. First, Tony Yaksh in the United States searched the brain by giving small localized injections to discover where morphine was acting to reduce pain. He found two areas, one in the midbrain, the other in the spinal cord where sensory messages were arriving from the tissues. Next, Kosterlitz in Aberdeen, as described before, ended a fifty-year search by discovering that the brain was itself producing its own narcoticlike substances. These endorphins, as they are known, were made by nerve cells that were particularly concentrated in the two target areas discovered by Yaksh. Finally, Snyder found that the brain also made special protein receptors that snapped up narcotic molecules and changed the excitability of the nerve cells on which they resided.

With this series of discoveries, it was possible to put together a most curious story. In the first place, the brain contains its own system for controlling the arrival of pain-producing messages. One part of this system exists as a barrier zone in the spinal cord where the sensory nerve fibers enter. The other part exists in the midbrain, from which region control orders descend into the spinal cord and further reduce the incoming message. When narcotics are given as a medicine, they penetrate the brain and stimulate the very system the brain uses to control its own sensory input. There was an immediate practical consequence of the discovery of the location of the pain control systems. Because the spinal cord was one site of action, and because it is simple to make a needle penetrate to the surface of the spinal cord, it was possible to apply morphine precisely to the site where it is carrying out its useful action. This has grown into the widespread use of epidural narcotics, where a strong analgesia can be produced in an area of the body with a dose ten times smaller than that needed if the whole body is treated by tablets or injections. Because this targeted dose is small, the side effects, produced by the effect of narcotics on distant parts of the body, do not occur.

Cannabis is another herbal remedy with a terrible social reputation. It is going through a surprising revival as a therapeutic analgesic, which repeats with a gap of twenty years the story just described of the emergence of narcotics from being drugs of social menace to ones of therapeutic value with rational understanding. Cannabis has been used for millennia as folk medicine for poorly defined problems and for social entertainment. Queen Victoria used tincture of cannabis for her menstrual cramps. In this century, patients began to report beneficial effects of low doses of cannabis in very specific conditions including nausea and pain from multiple sclerosis. In the hostile social atmosphere, where the use of cannabis and narcotics was equated with abuse by irresponsible people, these reports were dismissed or ignored.

In the meantime, scientists were at work with the traditional series of investigations like those into opium. Cannabis was purified by Marek-Marsel Meshulam in Jerusalem and found to contain a series of active compounds called cannabinoids. It was found that the brain itself was normally producing cannabislike compounds. Finally, to round off the progress that imitates the investigation of narcotics, special receptors tuned to react specifically to natural and synthetic cannabinoids were found to be widespread in the brain and body tissues. While cannabinoids are at present only used in legal practice to control certain types of vomiting, it seems highly likely that they will also emerge to control some pains.

Antidepressants

People in pain may well be depressed by their struggle and need treatment for their depression. However, the medicine used against depression have an action against pain that is completely separate from their action in depression. These medicines act by increasing the level in the brain of neurotransmitters, which are used to carry nerve impulses from one cell to another. One of these chemicals, called serotonin or 5-hydroxytryptamine, improves the mood. However, the same chemical is the transmitter used by one of the systems by which the brain controls the nerve impulses that arise in the spinal cord and signal injury to the brain. In this way, antidepressants can decrease the incoming signal from cord to brain and improve pain relief. They are used where the narcotics do not work because nerve damage has inactivated the inhibitory mechanisms that are normally activated by narcotics. They are used to treat the pains arising after shingles, which I mentioned earlier. But the effect is weak and these drugs only just pass rigorous tests as analgesics.

Antiepileptic Drugs

When groups of nerve cells become very excitable, they may all gang up and fire in synchrony. This is what happens to cells in the cerebral cortex during an epileptic fit. Antiepileptic drugs prevent this simultaneous firing. Trigeminal neuralgia, as described in the last chapter, is a horrible pain in the face that appears as sudden violent stabs. It seems likely that hyperexcitable cells in the brainstem, which receives the sensory nerves from the face, fire in synchrony to produce the stab of pain, and they too are prevented from firing together by antiepileptic drugs.

Overview of Medicines

Medicine has been described that acts on peripheral tissue to block some aspect of inflammation, while other drugs act on spinal cord and brain to block incoming messages. But all of them have side effects, which is hardly surprising since they are powerful and they interrupt a coordinated mechanism. Research aims to increase their analgesic efficiency and reduce their side effects. This ambition assumes that there is an isolated pain system, like a burglar alarm system, within the body that could be switched off, leaving the whole brain and body to get on with its normal business. This may be fantasy if the system that signals pain is part of an integrated system with other important business to perform if it is not occupied in its pain-producing business. We are left then with drugs against some pains, but all of them have their limitations, and there are some pains associated with destroyed nerves that fail to respond to any medicines. Surely we ought to be able to do better.

Surgery

Surgery Directed at Peripheral Nerves

It is, of course, possible to inject local anesthetics around peripheral nerves or around the sensory roots entering the spinal cord. These injections stop all nerve impulses traveling in that nerve and completely stop pain of peripheral origin. This highly satisfactory treatment can be used for only a day or so, however. The reason for the time limit is that these drugs, which are all children and grandchildren of the herbal origin compound cocaine, not only stop nerve impulses but also stop the transport of chemicals along fibers. This transport is necessary for feeding the nerve fibers; if it is stopped for more than a few days, the fibers die of starvation. Drugs may be invented that stop only the nerve impulses, and these would be candidates for long-term trials.

Cutting nerves can, of course, be performed by surgery. It is done in some cancer patients who have only a short time to live. If they survive longer, the nerves regenerate, the pain returns, and it is difficult to operate a second time. Noordenbros and I learned a salutary lesson in six cases that were astonishingly similar. All of them had partially cut across a nerve in the wrist, the median nerve, which supplies the thumb, neighboring fingers, and palm of the hand. This nerve is often damaged in accidents, particularly if the hand breaks a window. All of these patients had felt desperate burning pain and tenderness in their useless hand for more than six months and were not helped by any treatments. Hand surgeons have become very skilled in grafting across gaps in nerves by deliberately suturing many strands of a fine nerve from elsewhere in the body to bridge the gap. The nerve fibers regenerate across the bridge and grow on to make contact with the distant structures. This delicate reconstructive surgery works very well if it is done soon after an accident that has torn a gap in a nerve.

We persuaded the patients to allow the surgeons to cut out a length of the nerve where it had been damaged and to graft in new nerves so the nerve could reconstruct itself. The immediate effect of the operation was that the pain was completely gone but the hand was numb and paralyzed. Slowly, over six months, the nerves grew back and sensation and movement reappeared in the hand. The sad part of the story is that every patient developed exactly the same pain state they had suffered before the operation. One patient killed himself in disappointment.

The moral of the story is that you should not operate on peripheral nerves in this state. The pathology that was originally in the nerve had migrated centrally into the spinal cord, where angry nerve cells had become hyperexcitable and were the cause of the pain. Even cutting undamaged nerves or roots, while temporarily stopping the pain, can eventually set off a worse pain generated by spinal cord cells.

There are, however, operations that are commonly done on peripheral nerves, ganglia, or roots. One we have already described, to treat trigeminal neuralgia, involves destroying part of the sensory ganglion or root serving the face. Sometimes these patients suffer grim pains in their numb face, undoubtedly produced by the central migration of the effect of cutting nerves. However, the patient often has a pain-free period for months and years before the pain returns. Frankly, no one understands what is going on. Tiny lesions or gentle surgical manipulation of the ganglion or flooding the area around the ganglion with glycerol all produce long periods of relief. There are even successful cases reported in the literature where the operation was accidentally carried out on the wrong side of the face. All this is very mysterious and suggests some very fragile pathology. Needless to say, none of these many types of operation has ever been subjected to the rigorous test described at the beginning of this chapter.

The plot thickens when we examine other types of pain-relieving surgery on limbs. A condition called Morton's metatarsalgia is characterized by a small, very tender area on the sole of the foot. It is classically believed to be caused by the trapping of one of the fine nerves of the foot that dives down between the small long bones of the foot. The treatment is to expose the trapped nerve and cut it out. The Royal National Orthopaedic Hospital routinely examined the piece of tissue cut out and rarely found any nerve fibers at all! What is going on? Is it that the nerve is irrelevant but that there is some damaged tissue that reconstructs itself after surgery?

Trapped nerves are a favorite explanation for pains with both patients and surgeons. It all sounds like good old-fashioned mechanical sense. "Trapped" nerves are regularly liberated by surgical dissection of various nerves in the arm, from the carpel tunnel in the wrist to nerves at the elbow and passing over the ribs. Results are variable and evidence for nerve damage is rare. There is no doubt that the surgery is disturbing tissue that is often inflamed, but the reason for the results, if any, is a puzzle.

Failure to cure pain by therapy directed at the periphery has naturally encouraged surgeons to move centrally. The first obvious target is the dorsal root ganglia, which nestle in their body hole between each vertebra. In the very common cases of back and neck pain with localized areas of tenderness, Skyrme Rees in Australia began an attempt to destroy nerves close to the vertebrae, and these operations have become common. Cuts were made in the region in an attempt to sever the nerves coming from the painful regions. After an initial period lasting some years, when the results were thought brilliant, the method fell into disrepute because of declining success and the obvious variability of which structures were cut.

A much more controlled method was developed in Holland in which a thick needle was lowered with precise X-ray guidance to rest on one dorsal root ganglion. The needle contained electrical wires that allowed the tip to be heated to exactly 70° C (158° F) to burn a small area around the tip. It was found that a considerable number of patients were relieved of their pain for fair periods of time. However, the apparent rationale for this treatment has now fallen into doubt. A Dutch team seeking to determine the best temperature for the treatment dropped the tip temperature to 40° C and achieved equally satisfactory results. The reason for the crisis is that this temperature is only tolerably warm and does not destroy tissue. I interpret the results as an inadvertent placebo trial that shows that the hot tip is not the effective component of the trial. The alternative possibilities are either that the effect is the consequence of placebo suggestion or that the presence of the needle without heat is the cause.

The most major peripheral surgery for pain is the removal of a herniated disc, which pushes out from the soft cushion of tissue between the vertebrae. Some surgeons also carry out a bone graft to prevent movement of the vertebrae. This operation has been performed for more than seventy years and rose to great popularity, but doubts are increasing. The role of the protruding disc is not clear because the protrusion and the pain vary independently. A placebo trial of a treatment designed to dissolve the disc showed a very high rate of recovery after the injection of innocuous fluid under general anesthesia. The supposed proof that the disc was cutting motor nerves and causing paralysis is now doubted because the pain can cause muscle wasting by central effects. The formerly enthusiastic proponents of the operation at the University of Miami have given it up in favor of a rigorous rehabilitation program. Again, it is not at all clear if the beneficial results of the operation are due to suggestion or to some nonspecific disturbance of tissue in the region of the apparent origin of the pain.

Operations on the Spinal Cord and Brain

Patients and their surgeons may be so overwhelmed by the unremitting domination of serious pain that they seek a fundamental answer by surgery, aiming to cut completely the pathway that transmits the message or to destroy the hypothetical pain center. If the ventral-lateral quadrant of the spinal cord is cut on one side, a clear analgesia appears on the opposite side of the body below the level of the cut. This loss of pain lasts for weeks and months but is replaced by even more unpleasant pains as nerve cells ahead of the cut build up their excitability. This operation is now rarely done except in some cases where cancer patients have only a short time to live. The very large improvement in the understanding of narcotic therapy for cancer pain means this has replaced the surgical attempt to cut the messages off from their destination.

In intractable cases where pain fails to respond to narcotics, neurosurgeons have made lesions in many locations that were believed to be candidates for the "pain center" or for being major message-carrying bundles. The results are remarkably similar no matter where the lesion is made. Initially there is a gratifying relief from pain, but, within days, weeks or months, pain returns, often with additional unpleasant characteristics. Another target of attack has been in parts of the brain classically assigned to "thinking" in the old dualistic scheme. The result, as in the case of frontal lobotomy, is a patient who has considerable cognitive deficits, particularly in holding attention. These patients may appear somewhat better, even though they report that the pain is as bad as ever, because their attention, normally riveted by their pain, drifts away onto other topics. Patients and relatives do not consider this relief worth the price.

Legitimate research continues to locate cells in the brain that are firing off in abnormal patterns and creating false signals. Such cells exist in the spinal cord when motorcycle crashes rip the entering roots from the spinal cord, leaving a numb paralyzed arm and ongoing pain generated by the spontaneous discharge of spinal cord cells attempting to compensate for their loss of input. The problem with an operation to remove the spinal cord cells is that, after a period of relief, cells further down the line react to the loss of input created by the surgery and take up the pain-producing activity.

Other Therapeutic Methods

There remain huge numbers of other methods, many of ancient origin and almost none of which have been subjected to rigorous trial, that yet retain enthusiastic advocates who proclaim their effectiveness.

Whole Person Therapies

Classical Acupuncture

According to ancient Chinese medicine, health is created and preserved by the flow of yin and yang, which are conflicting energies distributed to all parts of the body in defined channels, the meridians, that peak at points. The flow of energy can, it is claimed, be adjusted by inserting needles into the channels at these points. Chinese acupuncture investigators were unable to find the channels and now state that the effect is caused by the stimulation of special sensory nerve fibers. Most people think that acupuncture arrived in the West in recent times once China was opened up after the Nixon agreements. In fact, acupuncture was well known in the West in Elizabethan times, beginning with a textbook by Ten Rhijne in 1683. The European enthusiasm for acupuncture faded until it was reintroduced in the eighteenth century by the French from Indochina. It again faded until the mid-nineteenth century, when Admiral Perry returned from Japan with a Japanese government delegation. Acupuncture arrived and faded four times in four centuries. Its use is fading again both here and in China. A therapy whose popularity fluctuates depends on social belief. Double-blind testing of which point to needle has shown no specificity.

However, there is a variation of acupuncture that ignores the classical points and needles only painful spots of the type described in fibromyalgia. Janet Travell, who was White House physician to President Kennedy and treated his painful back, had been following the established practice of injecting local anesthetics into these painful spots. She discovered that it was not necessary to inject the local anesthetic as the penetrating needle was enough by itself. The needling produces a stab of pain followed by relaxation of the taut muscle band, followed in turn by some general soreness and then by relief, which may last for days. We see here again, as was suspected in some of the reactions to surgery, that generalized damage in a painful area may be followed by relief.

Yoga

Yoga was overtly intended to change the subjects' attitude to themselves and to the world, and yet puts muscles, tendons, and joints in highly unusual positions. The Alexander technique has similar aims and effects.

Movement

Exercise

It is generally believed but not rigorously tested that exercise, even when it causes pain, shortens periods of pain.

Relaxation

Apparently contradictory to exercise, relaxation is intended to correct abnormal postures.

Deep Massage

It is intended to break tissue free from constriction.

Manipulation, Osteopathy, and Chiropractic

The most organized of complementary therapies, manipulation, osteopathy, and chiropractic are designed to force tissue in abnormal locations into a normal position. They use escalating force to achieve this end. Evidence is lacking that the abnormal location existed in the first place and that normality is restored. A large number of so-called trials have been carried out, but they are so far uniformly unsatisfactory. The most common problem if one therapy is compared with another is that the differences are so obvious to the patient. For example, chiropractors organized an elaborate trial in which randomly assigned patients received either chiropractic therapy or a course of hospital physiotherapy, and the chiropractic patients did better. But the chiropractic patients were given a series of treatments in the private offices of practitioners who knew their profession was on trial. The other patients went without fanfare to receive routine physiotherapy by unchallenged practitioners who gave fewer and shorter sessions. No wonder chiropractic appeared superior.

In another recent analysis of various types of physiotherapy matched against so-called placebos, the only significant result was that the longer the treatment lasted, the better the outcome. This result applied equally to the designated therapies and their placebos. Whatever the placebo component may be, there is no doubt that these therapies, like many of those we have considered, produce some tissue damage.

Stimulation

Cold

A cold spray of the type used on athletes produces dramatic but brief relief. It cannot be caused by cooling the deep tissue that is damaged. The provoked impulses arriving in the spinal cord set off a temporary inhibition of the cells transmitting the injury signals.

Heat

There is no doubt about the generally comforting, relaxing effect of overall heating, as witnessed in the Roman caldarium and the modern sauna. It is a standard treatment for anyone's anxiety. A quite different treatment is local heat. In the words of T. S. Eliot:

Here comes the nurse with the red hot poultice,

Slaps it on and takes no notice

The nurse may take no notice, but the patient certainly does and shifts attention. Attention distraction is the basis of counterstimulation. A third and quite different intention of heat therapy is to increase circulation and thereby speed up the end of inflammation. Ultrasound and microwave therapy were invented for this purpose to heat deep tissue. The use of ultrasound will be described in the next chapter as an impressive example of a proven placebo effect.

Gentle Massage

Delightful.

Transcutaneous Electrical Nerve Stimulation

When I had discovered that the input from large, low-threshold sensory afferent nerves reduced the response of animal spinal cord cells to an input from injury-detecting fibers, Sweet and I tried in Boston in 1967 to produce the same effect, first on ourselves and then on patients. We knew that a small electrical stimulus preferentially excited the large fibers, so we superimposed a gentle tingling sensation on top of the painful area. When this can be achieved, it produces considerable relief of the pain. We extended the method to direct stimulation of nerves and the spinal cord, but the most common use is to try to produce the effect by simple electrodes on the skin. It is no panacea, but it works.

Pain Control

Many pain control methods using medicine or surgery or alternative medicine exist. The precise reason for their action, or even whether they are effective at all, is frequently not known. For the patient who benefits, it doesn't matter why it works. For the future, we do need to understand how to achieve analgesia, particularly for those pains that we now fail to control. A crucial component in all analgesias, no matter how they work, is the patient's belief that it works. That is the subject of the next chapter.

The Placebo Response

If you have strong reason to expect a pain to disappear, it may disappear. This is called a placebo response. This topic is at the very heart of understanding pain, yet it seems so unlikely that it has been unpopular and has been seriously examined only recently. Pain is usually treated by one of the techniques discussed in the last chapter. However, if a narcotic is given, the placebo response adds to its effect. If the treatment is accompanied by great expectations but has no specific analgesic action, the decreased pain is entirely the placebo response.

Why Is It Unpopular?

Quackery

The word *placebo* was used by Geoffrey Chaucer as long ago as 1340. His use refers to a psalm that begins "Placebo domino in regione vivorum" (I will please the Lord in the land of the living). He uses the word in mockery because it was the first word in prayers for the dead that were said by priests and friars who pestered the populace for money to sing these prayers. The derisory use of the word is similar to the phrase *hocus-pocus,* which is derived from "Hoc est corpus" (This is the body), which are the first words of the Mass. By the seventeenth century, the word had been adopted by doctors for inactive medicines that greatly impressed their patients. In 1628, Burton writes in the *Anatomy of Melancholy*, "There is no virtue in some remedies but a strong conceit and opinion." In 1807, President Thomas Jefferson wrote in his diary: "One of the most successful physicians I have ever known has assured me that he used more bread pills, drops of coloured water and powders of hickory ash than of all other medicines put together. I consider this a pious fraud." Jefferson expresses here a strict division between fraudulent placebos and medicines that were believed at the time to act by a rational mechanism. That division continues to this day, with therapy being judged either a placebo or true.

A Tiresome and Expensive Artifact

As described in chapter 8, it is required by law that a new drug be proven superior to a placebo. This perpetuates the separation of true versus imagined. The placebo response was taken by the drug companies as a meaningless error to be dissected out and discarded while attention is monopolized by the powerful action of the therapy, which did not depend on what the patient thought about it. Some therapies, such as surgery, were assumed to be so powerful and dominant that it was not only unethical but ridiculous to test for a placebo component to explain the success of the therapy. This is dualism in practice, as therapy is assumed to readjust body mechanisms while mental processes are assumed to be irrelevant. In this atmosphere, it is not surprising that the placebo response was not studied until very recently. Similarly, it was not considered likely that a patient's response might be a combination of physical and mental process. This led to the very mention of a placebo trial being taken as a hostile questioning of the logic on which the therapy was based. This hostility is shared by enthusiasts both for academic and for complimentary medicine.

The Reality of the Senses

Everyone assesses their own sanity by cross-checking senses with objective reality and with what other people say. We have special words for mismatches, such as *hallucination, delusion, madness,* and *drunkenness.* We trust our senses. Pain appears to us as the sensation provoked by injury. A trusted, impressive physician prescribes the very latest analgesic for your pain, and the pain disappears. Later, you learn that you were a guinea pig in a trial and you were in fact given a blank tablet. You are angry, cheated, embarrassed, and shaken. I have responded to placebo trials, and I am always mortified and ashamed of myself. The pill could have had no action on the reality of my injury and yet my sensation changed.

Given these three reasons, it is no wonder that the placebo is an unpopular topic. Some physicians think that anyone who responds to a placebo did not have a "real" pain: they are wrong. Some physicians think that a placebo is the same as no treatment: they too are wrong. Some think that only weak-minded suggestible people in minor pain respond: they are wrong. Even physicians respond to placebos! The placebo response is a powerful and widespread phenomenon. Let us therefore examine it, using four examples.

Wisdom Tooth Extraction

If you have suffered this assault you will know that three things happen after the extraction: it hurts, your face swells up like that of a chipmunk, and you can't open your mouth. These are three signs of the now familiar inflammation response to damaged tissue: pain, swelling, and stabilization of the part by reflex action to prevent movement.

At the Eastman Dental Hospital in London, a team had been working on ways to help this troublesome condition. Ultrasound has been used for many years as a physiotherapy treatment for inflammation, with the rationale that the deep heat produced by the absorption of sound would hasten the end of the inflammatory process. Despite the fact that five of six trials of ultrasound on limbs had shown it to be no better than a placebo, they decided to try its effect after tooth extraction. They found it to be highly effective. Then, in a double-blind test, they massaged the face with the ultrasound machine in some patients when the machine was working and in others when the machine was turned off, unknown to the doctor or patient. There was no difference in the two groups. Next, they tested whether the massage was having the effect and trained the patients to massage themselves in the way the doctor moved the machine. This self-administered massage had no effect. Up to this point, they had shown a typical placebo effect in which a doctor in a white coat massaging the face with an impressive machine had a marked effect irrespective of whether the machine was turned on or not. Furthermore, there was no effect if the patient applied the machine. The pain was reduced by doctor-applied treatment. They went further. The swelling of the face was markedly reduced and the ability to open the mouth was improved. This placebo effect was the same as that of a substantial dose of an anti-inflammatory steroid.

The reason for choosing this example from many is that the effect was not only on the pain, which unthinking dualists would say is "only mental," but also on the swelling and on the muscle spasm. Evidently, this placebo response required a doctor in a white coat with an impressive machine, and this combination improved not only the patients' subjective report but two objective signs of inflammation that are usually assigned to mechanical body processes. The rational explanation is that the brain affects hormones, which in turn affect inflammation.

Angina

In the 1940s and fifties, before the days of coronary bypass surgery, an operation intended to improve the circulation of blood through the heart was carried out on many thousands of patients with angina. The method was to ligate two arteries below the sternum in the belief that new blood vessels would grow to bypass the block, helping the heart. The rationale for this important operation, which was successful, came to be doubted when the new irrigation of the heart could not be observed. Astonishingly, two groups of surgeons and physicians, one at Harvard and the other at the University of Pennsylvania, obtained ethical permission to carry out a placebo trial. In one group of patients, arteries were exposed and ligated in the approved fashion, while in the other group the arteries were exposed but not blocked. The observing physicians and the volunteer patients did not know who had the true operation and who had the sham. The majority of patients in both groups of patients showed great improvement in the amount of reported pain, in their walking distance, in their consumption of drugs, and, in some cases, in the shape of their electrocardiogram. This is a rare example of surgery being submitted to a placebo trial, and the improvement of both groups was maintained over six months of observation despite a general belief that placebos have only a brief fading action.

Headache

Remedies for headache prescribed by doctors have all passed the test of a placebo trial, carried out in the relatively calm atmosphere of university hospital outpatient departments. The trial reported here was done in a much more unusual setting in 1995 by a group of doctors and scientists from the U.S. National Institutes of Health in Bethesda, Maryland. All of us have had headaches, with which we coped as well as we could. This trial was carried out on thirty patients whose headache disturbed them so much that they had gone to a hospital emergency room. It is important to realize that these patients would have been unusually worried by their headache and would have been full of expectation that powerful medicine would be given. Everyone was given an injection: one-third were given an aspirinlike drug, ketoralac, one-third a narcotic, meperidine, and the other third received a saline injection, all double-blind. All three injections produced an identical reduction of pain. The reason for choosing this example of a placebo effect is to emphasize the effect of the patients' expectation in exaggerating the placebo effect. The two drugs had normally been tested on relaxed patients and were shown to be superior to placebos. In this case, where the patients had deliberately sought emergency treatment and had a high expectation, the placebo was as effective as the drugs. Clearly, we are dealing with a subtle but powerful effect.

Postoperative Pain

In the 1950s, Henry Beecher, the Harvard pioneer of pain studies already mentioned, and his colleague Louis Lasagna studied a placebo versus morphine for the treatment of postoperative pain. Each patient received two medications, morphine and a placebo, but the order of the two was varied. Half the patients received the placebo first and half received the morphine first. They found that those who received the morphine first responded very well to the second injection, which was a placebo. By contrast, those who received the placebo first responded rather poorly to the second injection, which was morphine. Evidently, the patients' expectation had been built up by their experience of the first trial.

Fifty years later, this crucial and surprising effect was studied in Turin by Fabrizio Benedetti's team. They were treating thirty-three patients who had been operated on for the removal of part of one lung, a famously painful type of surgery. The patients were treated for their pain for twelve to eighteen hours after their operation with intravenous buprenorphine, a powerful narcotic. Their pain was carefully monitored during this time, as was their lung function. One of the side effects of narcotics is to depress respiration, although the patients were quite unaware of this effect.

As shown long ago by Beecher and Lasagna, the pain responses to narcotics is surprisingly variable. Some patients achieve excellent pain relief with one small dose while others require repeated doses to get the same reduction of pain. The same applies to the side effects. After the Turin patients had experienced satisfactory narcotic-induced pain relief for twelve to eighteen hours, each was given placebo injections of saline. The results show clearly that those patients who responded to small doses of the narcotics also responded well to small doses of the placebo. Even more surprisingly, those patients who, unknown to themselves, responded with respiratory depression to the narcotics also showed respiratory depression with the placebo. Evidently, the placebo mimics the details of the experienced pain relief and the covert side effects. It is apparent that the placebo may be indistinguishable from the drug after the patient had experienced the drug's effect.

Characteristics of the Placebo Response

Although this book concentrates on pain, it is important to realize that the expectation of effect is a strong factor in many other conditions. Powerful placebo responses have plagued trials in many pathological conditions including asthma, cough, diabetes, ulcers, vomiting, multiple sclerosis, and Parkinson's disease. Trials of therapy affecting mood states, such as anxiety, depression, and insomnia, challenge the investigators' ability to separate the "true" action from the suggestion that the therapy ought to work. A coffee drinker secretly fed decaffeinated coffee not only perks up but acquires the finger tremor associated with caffeine.

For those appalled by how much they spend on alcohol, a recent trial may amuse. Subjects were first given a flavored alcoholic drink in a controlled laboratory setting. The next day the subjects were given the same flavored drink but without the alcohol, and many declared themselves intoxicated. This phenomenon means that street drug dealers can peddle any old white powder as heroine or crack cocaine to experienced addicts. There are, of course, subtle hints that allow the user to detect repeated placebos. I once gave a doctor's wife a standard morphine injection for the pain of her broken arm. After a few minutes, she informed me that this dose was completely inadequate. I then discovered that she was a narcotic addict who was highly experienced in huge doses.

Returning to the topic of pain, it will not surprise you that there have been many attempts to trivialize the unpopular placebo response. Some doctors believe that it could only affect weak pains but the facts show that very severe pain is also modulated, which is not surprising because those in agony grasp at any straw. Some believe that only "imaginary" pains respond, but in fact cancer pains and postoperative pains, which are anything but imaginary by any definition, may respond. There are those, rooted in rational therapy, who think that the placebo is the equivalent of doing nothing, but, as we have just seen, doing nothing for a postoperative patient results in an increase of pain whereas the placebo may be followed by a reduction. Finally, there is a common belief that the placebo response is transient and fades on repetition. Decreased response to repeated dosage is a common effect with medication and is included in the patient's expectation. However, a prolonged response to sham surgery lasting at least six months has just been described. In a recent experimental trial, the placebo response, if anything, increased with repetitive trials of the placebo where it had been suggested that the effect would be maintained.

Nocebos

If a beneficial response can follow suggestion, it is not unexpected that the same applies to unpleasant effects. This is called the nocebo ("I will harm") response. All packets of medicine contain a leaflet describing the compound and at the bottom, in very small print, there is a section entitled "adverse drug reactions." To choose an example at random, the leaflet accompanying a powerful medicine for the treatment of migraine reads: "Adverse Drug Reactions may include: nausea, dizziness, warm sensation, fatigue, dry mouth, somnolence, heaviness or pressure in throat, neck, limbs and chest, muscle pain, muscle weakness and pins and needles." If a double-blind trial is carried out, the patient is instructed, of course, on the expected beneficial effects but is also warned of the possible side effects. When the patients receives a placebo, they may exhibit the beneficial response but they will also respond with a variety of the side effects about which they have been warned. In the old days, when Britain still had colonial troops, the army regulations included a special section for these men. In the part for East African troops, it was a capital offense for a soldier to put a death curse on a comrade. That is the ultimate nocebo.

Animals

People tend to assume that the placebo phenomenon must be the result of some complex, sophisticated, cognitive process and must therefore be unique to humans. However, we are moving to the conclusion that the phenomenon has to do with expectations, and animals have expectations and learned predictions. Rodents have a highly developed one, learning through trials to avoid any food that makes them sick. This is why rats and mice are so difficult to control by poisoned bait. If a rat is given an injection of a small dose of apomorphine in a space with which he is familiar, he salivates, his hair stands on end, and he curls up, looking miserable for half an hour. Months later, if the same rat is put in the same space and injected with a dose of saline, he puts on an identical performance as though he was again injected with apomorphine. This is a learned nocebo response.

If a rabbit has experienced a series of small insulin injections that decrease the blood glucose and is then given a saline injection in the same conditions, the animal reacts by raising its blood glucose. The animal has learned to counteract the effects of the drug by raising its blood sugar. With a saline injection, it reacts as though it has received the insulin. This is neither nocebo or placebo but shows that even an animal learns to counteract the expected effect. The discovery of this type of preemptive reaction in animals led to the whole phenomenon being classed as simple Pavlovian conditioning, and it was suggested that the same explanation must apply to the human placebo response. As I will explain below, I think that label is misleading.

Culture, Learning, and Expectation

The placebo response is the fulfillment of an expectation. Expectations are learned by individuals, and if enough individuals share the same expectation it is called a culture. Young children have not had time to learn that people in white coats with horrible-tasting medicine and needles bring relief. Therefore, they do not respond to placebos in an adult fashion even though they have learned that "Mommy will kiss it better." Unfortunately, the belief in mommy's omnipotence fades and is replaced by the beliefs of the community in which they live: on the Zambesi, the belief may center on the shaking of bones; on the Seine, on the power of Vichy water; and on the Hudson River, on a psychoanalyst.

The precise nature of the effective placebo becomes finely tuned and is exploited by the pharmaceutical companies who take advantage of the details. A colored tablet with corners is superior to a white round tablet. The colors have been investigated. Red is associated with power, while green and blue calm. Capsules containing colored beads are superior to any tablet. As everyone "knows," when doctors are getting serious they give injections. A saline injection is superior to any tablet. When doctors are very, very serious, they give intravenous injections, so an intravenous injection of saline beats an intramuscular injection. A famous professor of medicine taught his students to give patients a tablet held in forceps with the explanation that it was too powerful to be touched by fingers.

Jefferson would condemn this cynicism as a "pious fraud." However, in the most common situation, the therapist truly believes in the power of the treatment and communicates this belief to the patient.

The first trials of a new therapy are famously effective. There is an air of excitement generated about a new therapy that is so infectious that even double-blind trials produce high benefit rates for both the therapy and the placebo. In an extreme example, psychiatrists gave tablets to anxious patients with an assurance that they were blank tablets. Many patients improved, and, on questioning, the patients volunteered their opinion that no one would be crazy enough to do such a thing and that the psychiatrists were secretly testing powerful medicine on the patients.

Society itself is changed by its belief in medicine and surgery. Pressures of time and money lead organizations to attempt to dissect out and deliver just the essentials of effective therapy. Treatment protocols are now common once the patient has been labeled with a single diagnosis. The patient embarks on a fixed treatment schedule with the same personal touch that is given to a dirty car entering a car wash. Complementary medicine retains a phase of personal tuned diagnosis and treatment to persuade the patients that they are individuals in need of overall personal adjustment. This effectively raises patients' expectations beyond their experience in orthodox medical practice, where they have been labeled as *lower back pain* and treated as such. Modern medical practice is very efficient in its aim to attack defined local disease, but that does not include the individuality of the person who or their expectations. In an anecdote which could be true, a senior psychiatrist and a young one are leaving work at the end of the day, and the young one says, "Are you not stressed out listening all day to the awful problems these people have?" The old one replies, "Who listens?!"

Personality

Given the unpopularity of the subject of the placebo, it is not surprising that the placebo response for pain was seen as a fault in a system that normally reported pain reliably in the presence of tissue damage. In the atmosphere of dualism, this proposed fault was located in the mental sphere. Clinical psychology was developed at the same time as the placebo began to be investigated, and clinical psychology concentrated on personality types. It was believed that placebo responders must have something wrong with their personalities. Were they hysterics, neurotics, overly suggestible, introspective, or what? Each of these traits was investigated repeatedly and no personality type was found diagnostic of the placebo responder.

This negative finding may not surprise you because we are beginning to relate the ability to respond to a placebo with the expectation of the subject. Expectation depends on learning the predictions of the culture and even more on learning from education and personal experience. Everyone is capable of learning, whatever their personality. We have seen that an individual may become a placebo responder after they have experienced a beneficial effect, so it is evident that a placebo response is not built in to the individual but is learned.

Depression and Expectation

Pain is inevitably depressing and the longer the pain continues, the deeper the depression. Pain so monopolizes attention that behavior and thinking are impoverished. Every action becomes an effort, including eating and talking. Even suicide eventually becomes too much of an effort. Poor devils in this state are full of melancholic, sad thoughts. They expect more of the same and foresee a miserable, deteriorating future. Because expectation is bleak, such people are poor placebo responders. Avoidpessimists if you are looking for placebo reactors.

Anxiety and Expectation

Anxiety is another form of altered thinking associated with pain and often alternating with depression. Whereas depression is a passive state of expectation in which the worst is inevitable and requires no action, anxiety is full of action. The anxious person is convinced that the future is threatening and that it demands active defence. Anxiety comes in two forms: in free-floating anxiety, everything is threatening; in fixed anxiety, a particular event is expected to be threatening. Almost all of us have suffered a fixed anxiety as an exam approached and have been deeply irritated by those classmates who cheerily breezed through exams. The sudden onset of an acute pain inevitably signals threat, rivets attention, triggers anxiety and demands action. The action chosen will depend on learned expectation. The expectation depends on the diagnosis and treatment, both of which are culturally determined. The victim may be highly experienced with the pain, such as menstrual cramps or migraine attacks, when their expectation at least includes eventual relief. But more commonly, pain is a surprise with urgent demands.

The worst-case scenario diagnosis depends on personal experience. A patient with an awful family history of early death from heart attacks reasonably fears each chest twinge as the onset of his fatal attack. Someone who has nursed a parent through a painful cancer reasonably fears each pain as a sign of the beginning of their own end. I admit with a sad nostalgia that I suffered a series of rapidly "fatal" diseases as a medical student molded by sharing the misery of patients. Self-diagnosis sets up the need for anxiety and reaction and is determined by experience. It is followed by seeking treatment that is believed to be appropriate and effective. Expectation of success is a crucial component. The response to treatment is associated with a decline of anxiety, but the placebo response is not simply the decrease of the anxiety as the two have quite different sizes and time courses. When treatment fails, everyone is in trouble. The patient may replace anxiety with depression. An alternative is to maintain the anxiety and the obsession that effective treatment must exist somewhere. The result is a continuous search through the healing professions to exhaust patient, pocket book and therapists. Here lies mutual anger, alienation and repeated disappointment. An alternative to this exhausting alternation of anxiety and depression and failed expectation is calm resignation with no expectation of relief, of which the rare individual is capable.

The placebo response is played out on the stage of expectation, which is created by the patients and their experience and culture, by the reputation of the therapy and by the attitude of the therapists.

Mechanisms

The existence of the powerful phenomenon of the placebo response is deeply disturbing to any classical picture of the nature of pain mechanisms but has finally led to serious study. Some possible solutions to the problem of the placebo were rapidly dismissed. The responder is not just lying about their feelings in order to satisfy the expectations of the therapist. There is no personality "type" who responds to placebos. The response does not depend on the relief of anxiety and stress, which in turn produces some pain relief. The pain relief is specific.

An important episode in the investigation of the placebo response to pain took place in 1978 in San Francisco with the work of Levine and Fields. It was known by then that the body contained its own endogenous narcotic system, and they wondered whether the placebo response resulted from the body mobilizing these endorphins. They gave subjects a narcotic antagonist, naloxone, and showed that their subjects could no longer reduce their tooth extraction pain with a placebo reaction. This very important observation has been repeated by others, and some find the same effect. These experiments had a huge effect on the understanding of placebos.

Until that time, explanations of the placebo response had been magical, mystical, and vague and were hidden in obscurantist terms with doubtful meanings such as *cognitive dissonance*. Suddenly, with these experiments, the subject emerged in the contemporary world of hard neuroscience with a defined pharmacology. The placebo became respectable. However, welcome as these results are, we should not be diverted from the question of what precisely is going on in the brain just because we think the brain uses one of the chemicals available to it to achieve an end that remains to be explained on a deeper level than merely identifying a particular chemical used as a neurotransmitter and modulator.

Returning to the fundamental nature of the placebo effect, the observations described above that animals can demonstrate a placebo or nocebo effect changed many investigators' way of thinking. Until that time, the placebo effect had been assumed to be a cognitive process, whereas animals had been denied the luxury of thinking from the time of Descartes. (This benighted opinion is not held by anglophone readers who have been raised on *Alice in Wonderland* and *The Wind in the Willows*.)

The best-studied form of animal learning is the Pavlovian conditioned response. If a dog is offered meat, it salivates. If a bell is rung every time the meat is presented, a dog begins to salivate if only the bell sounds. This is learning by association. It was assumed that the new response to the bell, called the conditioned response, was some automatic coupling in the brain that did not involve "thinking." It was therefore proposed that the placebo response was a similar type of conditioned response. The analysis of the placebo as a conditioned response went in the following steps. Pain is normally coupled with treatment and the response is relief. The treatment may normally be a doctor giving a pill. In the conditioned learning process, the relief becomes associated with the doctor and his pill. The specific content of the pill is not part of the learning process. Once the conditioned response has been established, the doctor can give a blank pill and produce the same response. So goes the hypothesis.

This paradigm was then repeated in humans, most carefully by Voudouris at La Trobe University in Australia. Volunteers first experienced a strong painful stimulus to one arm. Next, a cream was applied to the arm, which they were told was a powerful analgesic, although it was in fact a simple face cream. The previously painful shock was then applied again, but the strength of the shock had secretly been turned down so they felt only an innocuous tingle. At this stage, the subject had experienced a marked reduction of pain associated with the treatment. Finally, the stimulus was returned to its original painful strength and the innocuous cream again applied. Now the experienced subjects reported that the stimulus was weak! This very important, simple experiment has many consequences. It was shown in all the various combinations of stimulus and "treatment" that only those subjects who believed they had personally experienced the analgesic effectiveness of the cream then gave a placebo response. This was the first time that a placebo response had been deliberately produced in an experiment. It shows that the subject's experience of relief is the key.

This crucial experiment led to new thinking. In the first place, Pavlovian conditioning responses, which are apparent throughout the animal kingdom (even cockroaches can learn), came to be thought of as some entirely mechanical process. However, very careful repeated investigations in humans have never shown that conditioned responses are free of cognitive awareness. It would therefore be quite wrong to propose that if the placebo response in man is a conditioned response it is therefore equally wrong to assume that there is no cognitive component. *Pari passu*, it is wrong to assume with Descartes that animals are simply automata reacting second by second. Animals too have expectations of reward and punishment, as any pet owner knows.

The demolition of the proposal that the human placebo response was an automatic association of stimuli and response was shown in a very simple experiment by Montgomery and Kirsch in Connecticut. They precisely repeated and confirmed the work of Voudoris but added a new group of subjects. These people had been told that the stimulus was being deliberately lowered during the training session. None of these subjects developed a placebo response. The importance of this experiment is that these subjects had received precisely the same series of stimuli and reported the same responses to the training stimuli as the eventual placebo responders. If the conditioned response, the placebo reaction, had been simply the consequence of the physical stimuli and the reported responses, they too should have been conditioned. In fact, they did not become responders because their expectations had been more strongly influenced by the verbal information that the stimulus was going to be reduced. The creation of a placebo response depended on having the experience that the pain response intensity seemed in one case to be related to the presence of an "anesthetic" cream. In the other case, the placebo response did not develop when the subjects had also been told that the response intensity would drop because the stimulus intensity was reduced.

The amount of pain perceived corresponds both to the size of the stimulus and the amount of pain the subject believes it is appropriate to expect when the appropriate therapy appears to have been given.

All of this leaves us with a question, "What precisely is a placebo?" It can not be a stimulus because, by definition, it is completely inactive. If a placebo is given in complete secrecy, nothing happens. We have seen that the placebo response is linked to the patient's expectation. Part of the response of a patient to any therapy relates to the patient's expectation of a beneficial effect. There is therefore a placebo component in any therapy. This leads us to examine what is meant by "expectation." In the next chapter, I shall propose that pain occurs as the brain is analyzing the situation in terms of actions that might be appropriate. These actions are, in sequence, escape followed by guarding and then relief. In those terms, the placebo is not a stimulus but an action that experience has taught may be followed by relief.

Your Pain

The time has come to bring together all the phenomena we have discussed in the previous chapters and ask what precisely is going on in someone who senses pain. There is more to this than a conventional desire for neatness and synthesis because a profound understanding of one's own pain has itself a therapeutic effect and proposes a rationale for therapy. Furthermore, when our conscious awareness of pain starts and persists, a series of further processes is set in action so chronic pain is more than prolonged acute pain.

Attention

No conscious awareness of anything is possible until it has captured our attention. Our sense organs in the eyes, ears, nose, and body are in continuous action, day and night, whether we are awake or asleep. The central nervous system receives steady reports of all the events these sense organs are capable of detecting. Obviously, it would be a disaster of excess if we were continuously aware of the entire mass of arriving information. We completely ignore most of the information most of the time. And yet any fraction of this inflow is capable of riveting attention. For this to happen, there has to be a selective attention mechanism that must have a set of rules. Those rules are not arbitrary. Every species displays its rules, which incorporate a selection of those events that are important to its survival and well-being. Some rules seem to be built in. Large, sudden, novel occurrences have precedence in their attention-grabbing ability. I propose that the arrival in the nervous system of messages signaling tissue damage is another of these built-in high-priority events.

In some species the built-in selection can be very precise. At the London Zoo, Ronald Melzack arranged for a cut-out cardboard out-line of a hawk, with its characteristic short head and long tail, to be dragged on wires above ducklings raised in isolation. They froze and peeped alarm calls. To reset the apparatus, he dragged the cardboard cut-out backward across the ducklings. Now they saw a short tail and long neck and crackled with delight at the possible arrival of mother.

Obviously there is a learned component of our selective attention mechanism. The bored radar operator sits staring at the screen, which is a snow storm of random blinking dots. Let one of these dots begin to move in a consistent line and attention locks onto that dot to the exclusion of all others. Let a migraine sufferer detect a small twinkling area in the visual field and his attention is riveted on this trivial event because he has learned that the aura of his oncoming migraine attack begins with just such a scintillating area.

In social animals, subtle triggers of attention can be shared. In West Africa, two species of monkeys feed together in flocks but eat different fruits. Their main enemy is the monkey eagle. One of the species is quicker than the other to spot arriving eagles, so both species benefit from the alarm of one. In Australia, a grouse selects her ground nest close to a tree containing a hawk because the hawk's superior height and eyesight detects distant predators long before the earthbound grouse can. And so it is with humans, where attention is infectious.

The attention mechanism must be continuously scanning the available information in the incoming messages and assigning a prior-ity of biological importance. We have already described as an example of "thoughtless" decision the switch of attention in the car driver who is in conversation with a passenger while engaged in "unconscious" skilled driving, until some fool cuts in front of him, whereupon atten-tion promptly switches from conversation to avoidance. This brings out the second rule of selective attention, which is that only one tar-get at a time is permitted. Obviously, it is possible to switch attention back and forth quite rapidly. However, at any one instant, only one collection of information is available for conscious sensory analysis. This one object can be preset. An example is the detection of your name being mentioned in the random buzz of conversation at a cocktail party. Similarly, it is possible to scan a long list of names and detect the one you seek with no recall of any of the others.

It is not intuitively obvious that attention can be directed to only one subject at any one time. It would seem a rather ridiculous limitation in a mental process that clearly has freedom to rove over vast areas: "Shoes and ships and sealing-wax, cabbages and kings," as Lewis Carroll put it. An explanation for this strict limit on attention could be that sensory events are analyzed in terms of the action that might be appropriate to the event. If the aim of attention relates to appropriate action, then it follows that a fundamental requirement of nature is that only one action at a time is permitted. It is not possible to move forward and backward simultaneously. You must make up your mind. The explanation for the singularity of momentary attention would then derive from the purpose of attention, which is to assemble and highlight those aspects of the sensory input that would be relevant to carrying out one act.

Of course, rival sensory events may compete for attention. The tale of the ass who starves to death when placed equidistant between two bales of hay is indeed a myth that would never happen. Even real asses have a built-in requirement to make a choice. There may be many events occurring simultaneously, each of which demands attention. They are ordered by rank into a hierarchy in terms of biological importance. The practical consequence of this ordering was described repeatedly in the earlier chapters on the apparent paradox of the painless injury. Each of these victims was involved in a situation where some action, other than attending to their wound, had top priority. Getting out of a burning aircraft is more urgent than attending to a broken leg, for example. The attention does not oscillate between the two demands. One is assigned complete domination until safety is achieved. Only then is the alternative assigned the top position, whereupon attention shifts and pain occurs.

The workman in the course of a skilled task and the soccer player about to score a goal go on to complete the task with full attention despite the conflicting demands of their coincidental injury. Only when the conditions of the top priority fade is there a reassessment of the next most urgent priority. In conditions of complete "emergency analgesia," pain emerges as the dominant fact when the emergency is over. This priority ranking of importance is partly built in, partly learned from personal experience, and partly a component of culture.

Therapy that is based on a molding of attention is effective. It is called distraction. When a toddler trips, smacks into the pavement, and howls, what does a parent do? Pick it up, dance about, utter the inane "coo," "oo," and "ah," and kiss it better. These are distractions. Because you can attend to only one thing at a time, it also follows that you can have only one pain at a time. This fact led to many excellent folk remedies, such as hot poultices, horse linaments, and mustard plasters. They are called counterstimulants. When pain really sets in, attention is utterly monopolized and nothing else exists in the world but the pain. Many therapies attempt to intrude on this fixation. The distraction that is effective may be simple, but it will depend on established priorities. A game of cards, letting the cat out, or the sight of a hated neighbor can provide a brief interlude in pain. Some victims discover this for themselves and prolong their brief vacations from pain by inventing distractions; others get professional help in occupational therapy.

In another distraction therapy, given the pretentious title of cognitive therapy, the victim learns to daydream and play out an internal fantasy. It may be that they are on a warm sunny beach, or at a football game, or in their favorite bar. Some people can become very skilled at these distractions and give themselves longer and longer respites from their miserable pain.

Alerting Orientation and Exploration

As attention shifts to pain, alertness appears. There is something wrong. Alarm bells ring. Battle stations! Muscles tense and the body stiffens to a ramrod. Unknown to the victim, these overt changes are part of a massive reorganization of many parts of the body. The heart and vascular system get ready for action, the hormone system mobilizes sugar and alerts the immune system, the gut becomes stationary, and sleep as an option is canceled.

The eyes, head, and neck turn to inspect where the pain seems to be located. The hands explore the area. Muscles are contracted to learn what makes the pain worse and what eases it and to seek a comfortable position and then hold it. The end result is a body fixed in an overall pain posture. Muscles are in steady contraction and, as time goes by, some muscles grow while joints and tendons deteriorate because this frozen posture itself sets off local changes. The vascular and endocrine systems hold their emergency state if pain is prolonged, but these systems are not evolved to cope with this prolonged stress state. The quiet gut demonstrates its inactivity as constipation. Perhaps worst of all, sleep is impossible, and sufferers of chronic pain become completely exhausted. Even intermittent sleep deprivation drives the strongest of us into pretty peculiar ways of thinking, as any doctor on night duty and any parent with a newborn baby knows. Patients with chronic pain reach their wits' end as their grim experience is prolonged.

Clearly, this state of affairs needs therapeutic attack. The key word is *relaxation*, and much ingenuity has been used to attain it. The problem is to override a natural defence mechanism that has a protective role in brief emergencies but becomes maladaptive when prolonged. Drugs that inhibit the overactive muscle are commonly prescribed, but they are sedative and intellectually flattening. After a while, patients refuse them or become zombies. Physiotherapists have many ways of relaxing muscles and of reestablishing movement in frozen zones. First, they have to overcome the patients' natural belief that movement that produces pain does not necessarily increase the injury and that lack of movement that seems at first to prevent pain eventually acts to prolong pain. Yoga and the Alexander technique are examples of posture training. Relaxation is not easy and training methods are needed. One successful version, biofeedback training, provides the patient with an electronic indicator of the amount of contraction in a muscle and allows the patient to judge, second by second, her success in relaxation. The patient has to learn how to relax and how to prolong the effect into real life outside the training sessions. Sleep follows relaxation, but it may need help of its own, so no patients should resist tablets until they can sleep on their own.

The Sensation of Pain Itself

We are used to discussing sensation as the consequence of stimulation in a series of boxes: first, injury generates an announcement of its presence in sensory nerves; second, the attention mechanism selects the incoming message as worthy of entry; and third, the brain generates the sensation of pain. But we have to ask how the brain interprets the input. The classical theory is that the brain analyzes the sensory input to determine what has happened and presents the answer as a pure sensation. I propose an alternative theory: that the brain analyzes the input in terms of what action would be appropriate.

Let us explore these alternative theories because they have practical consequences for pain therapy. If the classical theory were true, the first action of the brain is to identify the nature of the events that generated the sensory input. This should produce the first sensation of injury as pure pain. The next stage of the classical theory is that different parts of the brain perceive the pure sensation and generate an assessment of affect: "Is the pure pain miserable, dangerous, frightening, and so on?" My first reaction, on introspection, is that I have never felt a pure pain. Pain for me arrives as a complete package. A particular pain is at the same time painful, miserable, disturbing, and so on. I have never heard a patient speak of pain isolated from its companion affect.

Because classical theory assigns different parts of the brain to the task of the primary sensory analysis and the task of adding affect, one would expect some disease to separate pain from misery. No such disease is known. During neurosurgical operations, very small areas of brain can be stimulated, some of which cause pain. There has never been a report of pain evoked that was not accompanied by fear, misery, or other strong affects. Finally, there are parts of brain, such as the primary sensory cortex, that have been classically assigned the role of primary sensory analysis studies, but these are often reported as silent when the subject reports pain. Even for the sympathetic pain on hearing of the death of a friend, the sensation is inseparable from the sadness and loneliness.

Instead, let us examine the alternative, which is that the brain analyzes its sensory input in terms of the possible action that would be appropriate to the event that triggered the whole process. There is in this absolutely no suggestion that any action need take place. Trained subjects and stoics may receive a clearly painful stimulus with no overt movement, even though they can later report the nature of the pain they felt. There are elaborate and extensive areas of our brain concerned with motor planning as distinct from actual motor movement. It is precisely these areas that are most obviously active when the brain is imaged in subjects who are in pain but quite stationary.

We have described previously the areas found to be active while the subjects feel pain. The first area of surprise to be reported was the anterior cingulate, which becomes active in any act of attention, which is exactly what is expected given the evidence that attention is a prerequisite of pain. The other areas consistently reported as active by many investigators are the premotor cortex, the frontal lobes, the basal ganglia, and the cerebellum. The past hundred years of neurology have assigned these areas a role in the preparation for skilled, planned movement.

Because I am proposing a quite new hypothesis here, we should explore widely to see if there are facts that support the possibility that sensory analysis is carried out in terms of motor action that would be appropriate to the input. Mimicry is a motor act that demonstrates the subject has detected a movement and proves the detection by imitating. The earliest sign that babies are perceiving complex visual events is their astonishing mimicry of facial expression, opening the mouth, smiling, and so on. Giacomo Rizzolatti at Parma in Italy has found single cells in the premotor cortex of monkeys that respond very successfully when the animal makes a particular movement, such as grasping with the fingers. What is astonishing is that the same cell responds when the animal sees someone else grasping an object with the fingers. The learning of birdsong from other birds has been shown to depend completely on the intactness of the learning birds' song-producing motor system. In the early work on speech by Noam Chomsky and

Morris Halle, some types of word recognition were shown to depend on the listener mimicking the sound of the word. They called this "analysis by synthesis."

The reason for describing these examples of mimicry is that they occur during learning while the overt motor movements are suppressed. There is every reason to believe that their central representation remains as a premotor plan. In the case of pain, the analogy would be that the overt defensive reaction to a noxious input observed in the baby is suppressed in the adult although the pattern of responses is retained as a possible reaction.

Could this new hypothesis be nothing but playing with words? It might not be possible to unravel the difference between the representation of the stimulus event itself and the representation of the likely motor response to the event. The most striking example of a clear difference comes from studies of the very elaborate auditory cortex of bats. Bats locate their flying food by emitting a series of "shouts" at moths and listening for returning echoes. The bat's cortex uses the time and sound of the echo to calculate the range, vectors, and flight path of the moth. Does the bat calculate the position of the moth when the echo bounced off it? That would result in classical sensation. In contrast, the bat collates all the data available to it and calculates the collision course it must take in order to arrive where the moth will be. The stimulus is analyzed in terms of optimal motor reaction. A great deal of study has been done to discover the mechanism for eye and head movement by which turning can bring a visual target directly in front of the eyes. When an animal is presented with a new visual target, do the brain cells respond to a signal where the target is in the visual world, which they would in a classical sensory system? The answer is no. The cells that respond are those involved in the decision of which muscles in the eyes and neck will bring the target in front of the eyes. The visual image location is analyzed in terms of the appropriate motor response.

The most astonishing example of the involvement of motor bias in sensory interpretation is seen in people who have suffered a stroke that has destroyed their inferior parietal cortex. This part of the brain lies on the side of the brain just above the pinna of the ear. If the stroke has occurred on the right side of the brain, these people appear completely unaware of anything on the left side of their world. They appear blind and deaf to anything occurring on the left and, most bizarre of all, when shown their own left hand they deny that it is part of them. When asked to draw the numbers on a clock face, they fill in the numbers from one to six and then stop.

Italian doctors in Milan showed that this neglect of the left half of the world even applied to the memory of a scene. They asked their patients to imagine that they were walking into the cathedral square in Milan by a road that enters the square opposite the cathedral on the south side. They were then asked to describe the buildings in the square. These citizens of Milan could recite the famous buildings on the east side of the square but were quite unable to recall any on the west. After a rest, they were asked to imagine that they were entering the square by a road on the north side. Now, imagining that they are facing south, they can recite the names of buildings to their right on the west side of the square but are quite unable to list the buildings on the other side, which they had been able to describe when they imagined they were facing the other way. This all sounds like complete madness, and it is true that the patients are still suffering from their recent stroke and usually paralyzed on the left side. However, this precise condition of one-sided neglect has been observed repeatedly in patients in many countries.

Classical theory explained this condition by proposing that there was a complete sensory map of the outside world and of the body in the brain and that the stroke had destroyed the left side of the map. Now comes the really astounding fact. Italian doctors, whose results were confirmed by many others, discovered that stimulation of the vestibular system in the ear completely restored all sensation on the left side. It disappeared again as soon as the stimulation stopped. What could be going on? The vestibules in the ear continually inform the motor system about the body's position in the up-down and sideways directions. It is our major organ of balance. It is obvious that the map had not been destroyed in the patients but that they did not have the ability to refer to the entire left side of the map. How could that be? Disturbed messages from the vestibular system, which controls sensory motor posture, had slammed the frame of reference for the whole brain so far to the right that it was unable to perform both its sensory and motor tasks on the left side. It is apparent that we can sense only those events to which we can make an appropriate motor response.

What would be the consequences of following the hypothesis that sensory events are analyzed in terms of the appropriate motor responses? It would provide a more satisfactory explanation of the paradoxes produced by the classical hypothesis and would help us begin to understand the facts just described. What are the appropriate motor responses to the arrival of injury signals? They attempt: first, to remove the stimulus; second, to adopt a posture to limit further injury and optimize recovery; and third, to seek safety, relief, and cure. The youngest, most inexperienced animal may attempt a series of these responses triggered by built-in mechanisms. As the animal grows in experience, the reactions will become more subtle, elaborate, and sophisticated. If the sequence is frustrated at any stage, the sensation and posture remain.

Humans develop and elaborate the three-stage responses from the moment of birth. Until some ten years ago, pain in newborn babies was neglected and even denied by professionals for two reasons. The first was that the human brain was seen as a hierarchy of levels: the spinal cord, the brainstem, and the cortex. This view had been introduced by Hughlings Jackson in the nineteenth century. Each level was believed to dominate and control the level below. The hierarchy of levels was believed to be an evolutionary development and to be repeated in the development of each individual. The ability to feel pain, misery, and suffering was assigned as a property unique to the cortex. All reactions to injury in the absence of cortex were called simple reflexes and were thought mechanical and free of sensation or emotion. This was the view that led Descartes to deny mind to lower creatures, and was perpetuated in post-Darwinian neurology, which assigned sensation and emotion to recently evolved structures such as the forebrain and cortex. It is true that we have a poorly developed cortex at birth. It takes two years for the major motor outflow from the cortex to establish control over the spinal cord. The second line of reasoning was that, because babies could not feel pain, there was no point in giving them potentially dangerous analgesic drugs.

Fortunately, thinking has changed such that pain in babies and children has become a major focus of attention. Of the many pioneers who brought about this revolution, I will mention two: K. J. S. Anand, a pediatric anesthesiologist of Sikh origin now living in the United States, and Maria Fitzgerald, a neuroscientist in London. Turning away from endless inconsequential philosophy on whether a baby feels pain, they and others turned to practical objective measures. The first question was whether a baby who must be operated on soon after birth would recover better if treated with a full battery of analgesics, as would be given to an adult. The answer was a powerful yes, and the result has been a marked change in neonatal anaesthesia and in survival. The second question was whether the injuries commonly suffered by babies, especially premature ones, produce a long-term shift of behavior. Again the answer is yes. Fitzgerald showed that even the act of taking a blood sample without anesthesia changed the motor behavior of premature babies.

This has focused new studies on long-term effects. Most surprising is a Swedish study, confirmed in Canada, in which a large group of boys who had been circumcised soon after birth without anesthesia were compared with similar boys who were not circumcised. They observed these children six months later when they received their standard immunization injections. The circumcised boys struggled, shouted, and cried far more than the others. Subtle controls showed that it was indeed the circumcision that had engendered the abnormal reaction to subsequent minor injury.

In the child and the adult, there is a continuous development of the way in which the victim moves through the three stages of reaction. Experience teaches skills. Society adds its methods of help and its prohibitions. Expectation becomes tuned.

Finally, we need to reexamine whether pain signals the presence of a stimulus or whether it signals the stage reached in a sequence of possible actions. The placebo phenomenon represents a profound challenge to these alternatives. The placebo, by definition, is not active and so cannot change the signal produced by the stimulus. It can hardly be categorized as a distraction of attention. Someone who has received placebo treatment for pain does not actively switch attention to some alternative target. On the contrary, they passively await the onset of the beneficial effect of the placebo while continuing the active monitoring of the level of pain. If, however, the sensation of pain is associated with a series of potential actions, such as remove the stimulus, change posture, and seek safety, then eventually the appropriate action is to apply therapy. If the person's experience has taught him that a particular action is followed by relief, then he responds if he believes the action has occurred. In this scheme of thinking, the placebo is not a stimulus but an appropriate action. As such, the placebo terminates and cancels the sense expressed in terms of possible action. Pain is then best seen as a need state, like hunger and thirst, which are terminated by a consummatory act.

When Pain Persists

The Disease Develops

In previous chapters, we have looked at various examples where damage to tissue is followed by inflammation. The quality of the pain and what to do about it changes. In postoperative pain, the initial acts of tissue damage were carried out under anesthesia, and the patient wakes up to sense only the later stages where the body attempts repair. In slow-onset diseases, such as arthritis, pain escalates as the disease process extends. Pain may grow in sudden jumps as in some cancer pains where the tumor has expanded into new territory and blocks the normal flow of blood, the contents of the intestines, urine, or nerve impulses. Intermittent pains grow with each episode.

Someone in their sixties or seventies walking uphill may be struck by a chest pain. Once you stop walking, the pain goes. This is angina of effort, an announcement by the heart that it can no longer pump enough blood around the body in response to the energy demand of walking uphill. As time goes by, the arteries continue to clog and their maximum blood flow drops. As this proceeds, the steepness of the hill that can be climbed drops, the amount of exercise that pain permits drops, and rest periods prolong. Eventually, if untreated, the angina forbids even standing up. These are expected reasons why pain may persist or escalate that we may be able to attack at the source. However, there are a series of rather different changes to accompany pain that we must now examine because they play an important role in pain intensity.

Fear and Anxiety

Anyone who senses an unexpected new pain and does not feel fear is not normal. There is a natural fear of the unknown in all of us and this is coupled with a fear of the consequent future. As part of the innate urge to explore, there is an immediate urge to know what is going on. We fear the cause and its meaning. When a patient goes to the doctor with bad pain and tenderness in the abdomen, the doctor may diagnose appendicitis, cancer, an ulcer, constipation, and so on. The patient may laugh with relief if the diagnosis is appendicitis because he has learned to believe that this is cured with a minor operation. This is a socially educated twentieth-century response because, two hundred years ago, it might have been the worst of all the diagnoses, as many patients with this disorder were in rapid decline and died within days.

Quite obviously, the amount of fear and the target of the fear will depend crucially on the person, her experience, and her situation. A middle-aged man from a family in which all the men died of heart attacks in their fifties and sixties has good reason to blanch with terror at the first twinge of chest pain. There are those with reason to fear cancer who develop an obsessed phobia and become crippled by their inability to accept medical assurance that they do not have cancer. The fear may become the disease.

Fear of consequences can be even more widespread, wild, and personally eccentric and so remain hidden to the witness. "Who is going to marry me now?" asked the Israeli officer with the amputated leg. "What a fool they will think of me to have let this happen," said the machine shop foreman with the amputated foot. Fears do not often relate to death but very frequently to the manner of the death. Fears relate to jobs, to sports, to sex, and all manner of personal needs. There is a type of macho tough guy who has "never had a day's illness in his life" yet falls apart at the seams with his first experience of pain and fear that breaks through his accustomed absolute self-control. There is every reason for each person to identify fears of cause or of consequence.

Fear generates anxiety and anxiety focuses the attention. The more attention is locked, the worse the pain. There is therefore a marked correlation between pain and anxiety. The anxiety here is not the free-floating variety with a feeling of general disquiet that something is wrong but cannot be identified. The anxiety of pain is generated by the unknown and grows worse as the pain persists and short-term expectations of relief fail to be fulfilled.

A major aim of therapy should therefore be to identify, understand, and treat the anxiety. This may need to start immediately after an accident. In any emergency room, a type of very distressed patient can be seen who is agitated by the scare of the rough and tumble of what he has just been through, although pain is the complaint. Of course, the pain should be treated, but he responds best if he also get help to calm down. Unfortunately, most chronic pain patients have settled into a rather steady state of fear and anxiety that becomes progressively harder and harder to shift. This by itself is justification for early treatment.

A very good example is the effect on postoperative pain produced by quite a brief talk with the anesthesiologist before the operation. The aim is to educate the patient with a step-by-step explanation of the stages to be expected. The explanation allows the patient to face the progressive stages of her recovery with familiarity and thus with less tension and anxiety. This points to the value of education in decreasing anxiety by illuminating the unknown. Well-designed programs for the relief of chronic pain teach as much as the patient wishes to understand about her own pain problems. It is always surprising to me what a revelation such education is to patients who have been carrying around a load of magical mumbo-jumbo myths, which nourish their anxiety. A crucial example that hinders recovery guided by physiotherapy is the myth that no movement is permitted that increases pain because such movement would increase the injury. This myth freezes the patient into narrower and narrower ranges of movement. Ignorance is never blessed. Any knowledge that brings patients into a clearer appreciation of their condition decreases their anxiety. It is for that reason that I wrote this book.

It is true that there is a well-recognized type of patient who sits in front of the doctor and says, in effect, "Cure me," with the unspoken coda that he is totally passive and expects curative action to be impressed on him by others. One's heart sinks, especially when you recognize that you are the twentieth doctor who has been invited to cure this patient. Given that I have presented pain as an active process involved with the brain's analysis of appropriate behavior, I would prefer to see the patient as an active member of his own treatment team. Anxiety has been a traditional subject for psychiatrists and psychologists, and it is most encouraging that they are beginning to apply their skills to the specific anxiety component of pain.

Failure and Depression

If pain persists and treatment fails, it is not surprising that depression sets in. Some patients plod sadly on, convinced that somewhere in the world there is a therapist with the answer. For some, it is a variation of the same answer but administered by a therapist with the right stuff. It is well known in affluent countries for people to have more than ten repeated operations on the same painful back. Surgeons are nothing if not high in confidence, and are not above hinting to the patient that they had been unlucky to encounter incompetent butchers before they reached the right one.

Early in his career, the Canadian neurosurgeon Wilder Penfield learned that his sister had a brain tumor and said, "She must be operated on by the best neurosurgeon in the world. Me!" Sometimes these charismatic fireworks are associated with success and sometimes not. The higher the patient scales the ladder of more and more distinguished therapists, the harder the fall. Frustration and anger are added to depression. Depression is a progressive certainty in a miserable future. Attention scans every detail of the pain to confirm that no change for the better has occurred and that it is in fact even worse than suspected. Every small change becomes a catastrophe for some.

This grim picture of anxiety and depression, phobia and fatalism is so commonly seen in chronic pain patients that there are those who claim that these conditions become the primary cause of pain, rather than being secondary to the pain that caused the anxiety and depression. Needless to say, this view is popular among doctors committed to some therapy that has failed a particular patient. Such doctors believe they have given the appropriate therapy to the patient and if the patient fails to respond it must be the fault of the patient. There are, of course, psychologists of the "mind over matter" psychosomatic school who are happy to support doctors who claim that the apparent body fault must be produced by faulty thinking if patients have failed to respond to therapy. One important school of behavior therapy believes that one can condition the patient out of his pain by ignoring any sign or word associated with pain and by rewarding and encouraging any sign or word associated with nonpainful activity. Of course, smart patients soon learn what the therapists want and shut up about their pains. They are considered successes.

I have not seen a scrap of convincing evidence that the mood and attitude create the pain. A recent new successful therapy by the surgeon Bultitude at St. Thomas's Hospital in London provides clear evidence that the pain drives the attitude. A rare urological disease, "flank pain with haematuria," is characterized by intense pain, no known pathology, and no known therapy. The patients are anxious, depressed, heavy users of narcotics, and are at their wits' end. The treatment consists of flushing the affected kidney under anesthesia with a specific nerve poison. The patients become pain free and at the same time their anxious depressed personalities return to the normal range. These are not anxious, depressed personalities liable to create or exaggerate kidney pain.

However, there is no doubt that the pain-produced anxiety, fear, depression, and obsession feeds back onto attention and posture and makes pain and living with pain harder to bear. Therefore, every effort made to treat these helps the patient. Rehabilitation programs focusing on education, movement, and relief of fear, depression, and anxiety do not cure pains but give the patients a freer lifestyle.

Coping

Some fortunate patients can learn to cope with their ongoing pain. Coping is not ignoring. In fact, it is the opposite. These people have learned to live with their pain in a realistic context. The pain persists but no longer demands emergency responses. It is not a catastrophe signaling impending annihilation. Patients obviously need help to reach the understanding that their life is not threatened. Coping is the beginning of a series of steps that give a sense of understanding and a type of control.

Charles Berde at Harvard, a leader of the new breed of pediatrician attending to pain in children, lists characteristics of young people in prolonged pain of neuropathic origin that is demolishing their lives. Such people tend to be depressed, anxious, stressed, are in wheelchairs or on crutches, are missing school, have a distorted body image, have eating disorders, and are in awful relationships with their parents and siblings.

This gruesome picture tends to contrast with children of the same age with an obvious disease, rheumatoid arthritis, who follow the example of fellow sufferers and learn to cope.

I recommend going to talk with a Second World War amputee who has been in severe pain for fifty years. Give him a chance to talk and he will precisely describe his pain now and the misery of the early days after his injury. Somehow, with the help of his comrades, he learned to ignore the fool doctors who dismissed his pain and managed to weave a life around it. He will also tell you that some of those comrades coped by killing themselves with bullets or booze. Coping is clearly a skill that may be learned with help. There is no chance of coping if attention is monopolized by fear, anxiety, and depression. There is no chance of coping while passively awaiting death or the invention of a cure. Coping is an active process directed at everything other than the pain itself. It needs inspiration and inspired help to live with pain.

This chapter has described what is going on in the person in pain. Pain requires attention, and distraction helps. Alertness, orientation, and attempts to ease the pain involve muscle contraction, and relaxation is of benefit. I propose that the sensation of pain itself is the consequence of our brain analyzing the situation in terms of what action would be appropriate. When pain persists, fear, anxiety, and depression rivet attention and make it more difficult to cope. The person in pain is locked in a syndrome, and therapy should be directed at every aspect of that syndrome. In the last chapter we will look at the attitude of other people to the person in pain.

11

Other People's Pain

Anyone in pain is locked in a struggle for relief. The rest of us have the option of approach or retreat. In this chapter we examine the alternatives.

First Aid

Good parents have the right attitude to their children. They always approach and touch, no matter what the problem. For adults faced with someone in trouble, there is a quandary. Do you have the confidence and competence to break the taboo of privacy? Does the social situation permit intrusion?

A friend was flying over wartime China when he was shot down. He parachuted down, broke his leg on landing, and lay in a ditch beside a road. Many peasants walked by with no signs of seeing or hearing him. Eventually a group picked him up, put him on a cart, and took him to the nearby village. He said he understood those who had walked by because the district was in the shambles of famine, where to take on the responsibility of one more being was a threat to all.

At the opposite social extreme, I came across a crowd standing round a man who had been knocked down on a London street. They stood silently watching as blood spread over the road from a cut artery on his head. I pushed in and jammed my thumb on the artery. After some minutes, the unconscious man opened his eyes and said, in deep

Irish, "I've had a few jars" (pints of beer). His comrade, leaning on a lamppost, said, "He's right. We have had a few jars." The crowd began to laugh and to repeat what the two men had said. I understood their laughter. They had been standing in frozen awe, feeling incompetent to act. Their sense of guilt began to evaporate as the man woke up and declared his drunkenness. The police and ambulance men arrived and looked puzzled by the crowd's hilarity. These two episodes encapsulate the question of the appropriateness of reaction.

Many of us are brought up shielded from other people's miseries and drilled in respect for their sacrosanct privacy. At the same time, we are taught not to complain about our own miseries. I still suffer from that upbringing because it is not a good starting point in seeking help for oneself or for others. I am well aware that there are other cultures where complaint has been raised to a high art form. The problem of continued kvetching complaints, audible a block away, is that others come to judge them more for their form than for their content.

If you wish to see real professionals give first aid for pain, watch a football match. When a player is injured, the medics rush in. Their main weapons are massage and cooling. Pain-relieving sprays work only by cooling. Ice packs are applied and, for those without elaborate equipment, a bag of frozen peas from the freezer works wonderfully. It's all common sense and folk medicine, but it requires a degree of confidence that comes from experience and learning. It is a mystery to me why first aid is not taught to everyone in school. Biology is taught and children end up able to name dinosaurs and identify pistils, stamens, and anthers on a flower, but they cannot tell you what blood is for or why they pee. The aim would not be to produce a gang of overconfident doctors: we have enough of those. It would be sufficient if we grew up with an intrigued awareness of our own bodies in health and an unflinching willingness to offer a hand to others in trouble.

Caregivers

Those who care for someone in pain are involved in a sequence of reactions that can stretch out for very long periods. It begins with fear, which can grow into terror and a sense of catastrophe. Fear is infectious. At first there may be anxiety about the pain and its meaning and consequences. This anxiety can generalize into the caregiver being anxious about everything and can show itself as agitation. We have seen the way the initial period of vigorous escape melds into a quieter period in which the victim guards the painful area and avoids movement. This can be a time of inactivity, loginess, loss of appetite, and a desire to be left alone. After operations or accidents, this phase can continue, even when the pain has gone, into a state where the patient feels flattened, with no energy and a feeling of apathy. This time had not been well studied until recently. It was believed to be the long period of healing and was certainly marked by surprise and frustration when the patients longed to get up and resume their normal activities. This stage of apathy certainly irritates the patients and their caregivers when expectation predicts recovery but it does not come. The condition is similar to chronic fatigue syndrome, although that state is not usually preceded by dramatic illness. Exercise seems to make the condition worse, and the patient is faced with a prolonged, irritating period of passivity with the fading hope that it will eventually go, although it normally does. Self-limiting and puzzling periods of stillness have in the past attracted little sympathy or attention from doctors and are trying times for patient and caregiver.

Professor Hall and his team in the Anesthetic Department at St. George's Medical School in London recently began a special study of the phenomenon. They compared the course of recovery of patients who had had a hip replacement with those who had had an abdominal operation. Hip replacement these days may be thought of as a routine operation, but, in fact, it involves first the exposure of the hip joint, which is buried deep in the mass of muscle making up our upper leg, and then sawing off the top of the large femur bone and its joint, before replacing it with a plastic and metal prosthesis. By contrast, operations on the abdomen may seem minor in terms of the amount of tissue disturbed by the surgery. The abdominal wall, which is quite thin, is cut open to expose the viscera. These organs in the abdomen and pelvis have a limited supply of sensory nerves. The diseased parts are located and removed and the abdominal wall is sewn up. The speed of eventual full recovery was compared in the two types of operation and a striking difference was observed, with the hip replacement patient bouncing back far more quickly than those with abdominal operations. It is clear that the amount of damaged tissue cannot be the factor that leads to the prolonged exhaustion.

Patients who will have a hip replacement have usually gone through a very long period of developing distress resulting from the slow onset of osteoarthritis. It begins with pain in the upper leg on movement and with aching at the end of the day. As it develops, movement becomes more and more limited and there is extreme difficulty in walking and a nagging pain at rest that begins to disturb sleep. Analgesics help to ease the pain in the early stages but have little effect on the movement, and eventually these patients become crippled and exhausted by their struggle and lack of sleep. They have a very positive attitude to the operation and have good reason to look forward to relief from pain and a return of movement. Many have already experienced the operation on their other hip, and most will have witnessed their fellow suffers make a splendid recovery.

By extreme contrast, people who have their abdomens opened and explored by surgeons have a far more worrying and doubt-provoking experience. They have suffered growing vague discomfort, sometimes not even frankly painful, seeming to originate from somewhere in the depths of their mysterious bodies. They feel unwell and frightened by the very vagueness of what is wrong and what the prognosis might be. Doctors peering into one's various orifices can be a shaking invasion of one's guarded personal space. Even more so, surgeons rooting around among one's viscera is ultimately frightening by itself, and who knows what they found or missed?

The St. George's group believes that it must be the fundamental difference in patients' attitudes that explains the striking difference of their long-term postoperative recovery time. They may be right, but one must not forget that the nature of the tissue that has been disturbed by the operation is very different. It could be that the body's recovery and defense mechanisms are very different when muscles and joints are damaged than when organs within our bodies suffer. It is crucial that we accept this phenomenon of malaise as a problem to be understood and controlled rather than ignored as it has been in the past. Apathy in growing children is famous for driving parents to distraction when their offspring exhibit contemptuous boredom with the very events that lit up the parents in their youth. Caring for a friend who sinks into a torpor of apathy and abandons former pleasures can be equally irritating.

Caregivers must take an attitude to a form of intense activity exhibited by some pain sufferers: the search for cure. This can become an obsession, with the patient being consumed with certainty that somewhere someone has the complete answer. When doctors are involved, this state can move from the sad to the frankly dangerous. The insistence of a pitiful wreck of a patient puts tremendous pressure on physicians and surgeons. Desperation breeds desperate measures. An escalation of invasion can result in the repetition of failed operations and the use of untested, dramatic new procedures. The Sloane-Kettering cancer hospital in New York believes that 23 percent of the pain problems they witness are caused by the therapy. It has been one of the clear advantages of the pain clinics to be discussed later that they have protected patients from excessive overenthusiastic therapy.

A more gentle and innocuous response of patients is to turn to the wide variety of complementary medicine after their physicians and surgeons have failed to relieve their pain. No matter how crazy the theory, these practices flourish. In chapter 9, I promoted the merits of the placebo. The effective alternative therapists have a number of advantages on their side. They can give more time to the patient than the overstretched health professionals. They often maintain an exuberant enthusiasm that has faded in their jaded academic colleagues. They can therefore offer more warmth, optimism, and attention to patients who are beginning to feel lonely and abandoned. The younger alternative therapists tend to be lovely people, whereas cynicism can settle on their seniors.

When pain persists, it is almost inevitable that depression is added to anxiety. With no end in sight and the progressive decrease of possible activity, the patients naturally turn in on themselves. They are sad, and it becomes more difficult for others to cope with the sadness. The patient becomes crotchety, particularly with the well-meaning efforts of others to get them out of their shells. This cycle leads to a sense of loneliness and alienation. "Why me?" they ask angrily. "No one cares." 'They don't believe I am in pain." "What is going to happen to me?" These descents into melancholy need professional treatment. In the meantime, those who care are themselves pushed to desperation and even anger and retreat. The caregivers face burnout and need relief, a break, and a community to share their problems.

I write as a warning to caregivers about what may happen, not about what inevitably happens. There are those in pain and their friends who have a built-in genius for coping. Their secret is never to deny the pain or its consequences but face both with reality. They observe and experiment with what makes it better, with no expectations of miraculous cure. They become expert at spotting diversions that give brief relief and then expand these periods of distraction until they become a way of life.

The Professionals

Medical Schools

Medical, dental, and veterinary schools were set up to purvey the principles of their profession. The cause of disease and fundamental cures were the main target. Symptoms such as pain were mere signposts on the road to the main aim. Symptom control was historically not worthy of the attention of serious men so this task was assigned to denizens of the depths of the hierarchy, such as nurses and physiotherapists. Even dying was not a worthy subject because, by general medical agreement, there was nothing more to be done.

In the past twenty-five years, pain has regained a status worth consideration in these professions. However, there are new powerful, justified claims on the strictly limited time available for medical education. The hugely tempting developments of molecular biology, biochemistry, and genetics have considerable new claims on the time of the students and on the interests of the faculty. To make time, traditional subjects such as gross anatomy have been condensed to a wizened nugget of the former two-year absorption of medical students in the dissecting room. Despite the obvious fact that pain is the most common complaint and the reason why patients visit their doctors, the subject as such has made little progress in capturing jealously guarded class time. In the preclinical years, pain can be "explained" in fifteen minutes by mouthing the hundred-year-old myths that there are pain fibers in the peripheral nerves and a pain tract in the spinal cord with a pain center in the thalamus. A few hours of lecture have been inserted to cover the whole of psychology. The pharmacologist may give a one-hour lecture on analgesics. In the clinical years there may be just a single session on pain. This means that the fully qualified doctor usually emerges with only three to four hours of tuition on pain.

Deans and professors ritualistically bemoan the lack of time devoted to pain, but there has been little progress. It is true that elective courses on pain appear and are attended with enthusiasm by psychologists and pharmacologists as well as medical students. If the table is so bare in the medical school, it is not surprising that nurses, physiotherapists, occupational therapists, and psychologists fare little better in their need to understand the subject that will take up so much of their working life.

The Specialists

In most countries, the main specialities in medicine and surgery include no training requirements on pain management. However, the old neglect may be on its way out. In Britain, psychiatrists have just recognized pain as a worthwhile topic. But anesthesiologists from all over the world have taken up the challenge in a very hopeful way. I know of no other profession in any field that has twice succeeded in doing exactly what they set out to do. In their first fifty years, they aimed to provide pain-free operations and, by the turn of the century, had invented and applied several types of general anesthesia and produced local and regional anesthesia. For their second success, they tackled the problem that they were undoubtedly killing some of the patients with their anesthesia. Within another fifty years, they had made anesthesia astonishingly safe by applying great skill and ingenuity. Now they have refined anesthesia by training the controller to eliminate accidents and errors and to introduce methods while monitoring the patient's condition breath by breath so that anyone, from early premature babies to centenarians, can be assured of safe anesthesia.

Their old masters, the surgeons, noticing that they seemed to have the patients under remarkable control, began surrendering more responsibility. Anesthesiologists should now keep the patients alive during the operation and see that they recover afterward. This led the anesthesiologists to take charge of the recovery room and move from there into intensive care units, which are more or less the same. Now they were faced with the major problem of pain in conscious patients, and from there it was only one more step to patients in prolonged pain and pain clinics. They now play the leading role in the clinical fight against pain, but the problem is too much for the time and skill of one specialty and the involvement of more is badly needed.

Organizations

The picture is not pretty in any country. In what follows, I will use data mainly from Britain, which is a middling developed country with a fifty-year-old comprehensive national health service, although about 20 percent of the population also uses private medicine on some occasions. Britain spends 6 percent of its gross national product on health, while France spends 8 percent, Germany 9 percent, and the United States 12 percent.

Acute Pain Teams

Obviously, everyone has a fully trained anesthesiologist present during surgery. Some 80 percent of patients are visited for between five and twenty minutes before the operation by the anesthesiologist for examination, reassurance, and familiarization, but 20 percent first see their masked anesthesiologist in surgery. In 1997, the government audited the state of pain control. The Royal Colleges of Surgery and of Anaesthetics had declared that "failure to relieve pain is morally and ethically unacceptable." The government survey found that 50 percent of patients in hospital had avoidable pain, largely attributable to disorganized guidelines for the staff. Financial pressures increase the frequency of day surgery and it was found that 10 to 20 percent of patients had quite unacceptable pain at home as a result of poorly planned relief. The newer methods of pain control require equipment, such as patient-controlled analgesia, or skilled staff attention, such as epidural anesthesia, but these were available for only a small minority. In order to provide a pain control service on the wards, for accident victims and for crises, hospitals have organized pain control teams that are ready to move to any patients. However, only 58 percent of hospitals had such a team.

Pain Clinics

John Bonica in Seattle in the 1950s invented the idea of a pain clinic. The idea was that patients referred by other doctors would be seen by a board of experienced pain doctors. Too often, in any country, patients make their own diagnosis and seek one specialist who might be an orthopedic surgeon or a gynecologist or a neurosurgeon. The chosen specialist then does his thing and, if it fails to work, the patient seeks another specialist. This process is clearly expensive, time-wasting, and potentially dangerous. The ideal pain clinic was to eliminate "doctor shopping," to present the patient to a variety of specialists with a concern for pain problems, and to permit a coordinated treatment plan that could include psychotherapy and physiotherapy. Although this was the aim, it soon became apparent that patients seen in a pain clinic had already seen other specialists who had failed. Furthermore, providing access to many pain specialists was very expensive or impossible.

In Britain, there are some very impressive pain clinics that fulfilled the original ideal. The Royal College of Anaesthetists made it compulsory for training anesthesiologists to spend time in a pain clinic. The result is that 87 percent of hospitals have something called a pain clinic, but 30 percent of these are staffed by a single anesthesiologist, 40 percent have one anesthesiologist and one aide, and only 20 percent approach the multidisciplinary ideal. There is a great need for these understaffed clinics and so there are very long waiting lists. The question now arises of the cost-effectiveness of these less than ideal clinics. The evidence is good that they save money because they decrease the costs of very expensive surgery, drugs, visits to doctors, and home care.

Hospices

The origin of the modern hospice movement can be attributed to Dame Cicely Saunders. She had learned as a young woman that care of the dying was left entirely in the hands of immensely kind-hearted people who had no training and no support from doctors. The lack of pain control was the most obvious tragedy. By the 1960s, she had set up St. Christopher's Hospice in London with a staff dedicated to the patients and to pragmatic research on how to ease their misery. It has been a wonderful success story, achieving particularly good pain control without reducing the patients to zombies. It has been widely imitated all over the world, with variations including resident hospices, palliative care wards in hospitals, and maintenance at home with visiting expert teams.

Rehabilitation Units

Some patients, most commonly those with back pain, have been in pain for years and have not responded to any of the conventional treatments. There are large numbers of such people hidden away, a perpetual misery to themselves and to those who care for them. Ten years ago, a program was developed for just such people at St. Thomas's Hospital in London. Groups of ten to twenty people are invited to take part in a program lasting three to four weeks in the hospital. Patients are selected who have been in disabling pain for more than a year, have had complete diagnostic studies, and have not been substantially helped by any of the standard treatments. They enter as a group and organize to look after themselves in a nursing home. For four to five days a week, they become involved in progressive guided programs where they learn about pain and their thinking, decrease the amount of medicine they take, and relax better. The immediate outcome and six-month follow-up has been studied in more than a thousand patients. The aim is carefully explained to them. It is not to abolish their pain but to make them sufficiently confident to liberate them from immobility and from dependence on medicines and on others. It works. By the end of the course, they are walking faster over longer distances, climbing more stairs, and taking less medicine.

It is not at all clear which component of this complex mix is the effective one, and it may well be that the mixture combines with synergy to liberate them. It may be that the group itself is therapeutic with its infectious interactive mimicry. The long-term cost-effectiveness of this labor-intensive and therefore costly course is apparent to the patients' doctors, as they use less medical care and experience more independence. The goal of complete pain relief and a return to work is a completely unrealistic outcome measure that may occur but that hides the very real improvement in the quality of life experienced by most of these chronic pain patients.

Variations on this type of treatment have spread widely around the world as teams explore the practical possibilities of how to help. The full-time inpatients' course may seem prohibitively expensive and disruptive, so outpatient visits are used instead. These schemes vary from daily highly programed courses to less and less frequent applications of fewer and fewer varieties of therapy. Eventually these programs become so diluted that they differ little from the old boring routines and are patently ineffective.

Self-Help Patients' Groups

The situation I have described leaves a scattered, invisible, underground population of people seriously disabled by their ongoing pain and yet abandoned and ignored. It is not surprising, in this situation, that self-help groups spring up all over the world as patients revolt at their neglect. A typical example was one called SHIP (Self-Help in Pain), founded by Rosalie Everett, a former nurse who had wrecked her back while lifting patients. Realizing that even good-hearted conscientious doctors had little to offer her, and having made her way through the smorgasbord of complementary therapies, including the Alexander technique, she realized that her experience had made her something of an expert. She was determined to share her experience with others and, with the help of local doctors, began to invite fellow sufferers to small group meetings. Such groups could easily turn into moaning misery sessions, where each patient competed with the others to determine who suffered most. Not with Rosalie Everett! She is a positive woman, determined to share useful resources and to offer mutual support. Lecturers were invited, visits were organized, and a community who knew each other was established. A telephone hotline was set up, manned by an experienced person who could talk to people in crisis. A newsletter was published with gleanings from the medical press, government regulations, pension advice, and so on. These groups now appear and flourish in many districts and countries. I am in awe of them. Their success and frequency proves their need. They remind the rest of us how far we still have to go.

The Pharmaceutical Industry

By the turn of the century, large pharmaceutical companies began to appear, often spinning off from chemical companies. Now, with growth and mergers, giants dominate. In their beginnings, innovative research with the best brains in the universities established the new pharmacology. In the nineteenth century, Robert Koch proposed that dyes that stained bacteria might kill them. The dye companies set to work, including I. G. Farben (*Farben* is German for "color"), the traditional fabric dyers in Basel, CIBA, Sandoz, and Hoffman-La Roche. Claude Bernard proposed that plant poisons were clues to important body processes. Henry Dale at the Wellcome Company set about on the analysis of ergot, a fungal infection of wheat that set off lethal epidemics when people ate bread containing "rusty" wheat, which produced St. Anthony's Fire and Dancing Madness, St. Vitus's Dance. From this he extracted and analyzed the major classes of compounds that are neurotransmitters. Brown-Sequard pointed out that the glands and the endocrine system are a source of body control, and the Parke-Davis Company isolated adrenaline.

These heady days of adventurous exploration and innovative research have faded from the drug company scene. Hard-headed accountants have taken control and, while profits have soared, fundamental research has shrunk. It is sad to scan the annual reports of these giants and note that the advertising and marketing budgets far outstrip the research budgets. Even the research departments are consumed with the scramble to invent "me too" varieties of drugs marketed by rivals. Caution along with an analysis of sales potential dominates planning. A new aspirin would be ideal, but the gamble to find a treatment for a relatively rare condition would not. A new treatment for diabetes would be a goldmine, but the painful neuropathy of diabetics is too uncommon and short lasting to be worthy of major investment. Only two major advances have appeared in the treatment of pain from the major drug companies in the past fifty years and they are both coincidental side effects: some antiepileptic drugs are effective in neuropathic pain and some antidepressant drugs are usefully analgesic.

But the picture is not all grim. While the old giants play a very cautious game, some of the new biotechnology companies are full of vigor. They are in intimate contact with the frontier of basic research and contribute by making astonishing new tools. Effects on pain have yet to appear, but these companies are hugely inventive and intelligent.

Society

Given the slow start of the professions in facing the challenge of pain, it is not surprising that most governments have trailed behind. The largest medical research center in the world is the National Institutes of Health in Bethesda, Maryland. The size of at least ten medical schools, it contains institutes dedicated to the major conditions such as cancer, heart disease, and so on but not even a section concerned with pain. The French national medical research organization has one excellent unit on pain problems, the Germans have two, but the British have none.

Charities are crucial in the support of medical research. In Britain, the Wellcome Trust alone funds as much research as the government's Medical Research Council, and the cancer charities support more cancer research than the government. In any country, a multiplicity of charities finance research and the well-being of those with many types of illness, including some with very obscure diseases. Yet no country in the world has a major charity devoted to those in pain. Why is that

so? It could be that people wish to see their money spent on fundamental cures, not on symptoms. There are societies that reasonably
seek a cure for arthritis, knowing that if they succeed the pain will go.
The Multiple Sclerosis Society does not divert funds to determine why
those who suffer multiple sclerosis are in pain. The International
Spinal Research Trust has in its charter that money may be used only
for research on the regeneration of nerves in the spinal cord and may
not be applied to symptom relief. There are headache and migraine
societies, but headaches could be considered self-contained entities.

This insistence on fundamental cure may be a partial explanation
for the absence of pain charities, but I believe that it cannot be complete. After all, there are many excellent, powerful charities for the
blind, the deaf, or amputees with the side intention of enriching their
daily lives and with no talk of cure or restoration. I suspect that the
entire subject of pain encompasses one of the last taboos. It is not a
topic of easy conversation. Better to speak of something else that offers
a chance of control. I have written this entire book with trepidation.
Has it skimmed over an abyss of dark horror that hides a terrible
threat? Presumably the reader who has reached this far has found some
method of coping with their own distaste of so disturbing a topic. One
may read about cancer from a psychologically isolated refuge even if
you have cancer, as I do. When I see someone in pain, I confess that I
still react with horror and would prefer to retreat. My response is the
occupational therapy of working on the topic. I do not believe one can
ever be familiar with pain. It is too deep.

Society is not kind to people in pain. Fifty million Americans are
partly or totally disabled for periods ranging from a few days to weeks
or months. Some are permanently disabled. A significant proportion
of chronic pain problems relate to the lower back. Some 60 percent of
the British population take more than a week off work for back pain
during their working life. In a telephone survey of 1,254 adult Americans, 56 percent reported some back pain in the preceeding year with
3 percent reporting lower back pain for more than a month. Surveys
of this type have been carried out in many countries and always show
the presence of very large numbers of people in trouble with pains, of
which back pains, headaches, and arthritis are the most common. The

fact that a proportion are suffering from very prolonged episodes means that available treatment must be ineffective.

One might think that such a vast problem would be a subject worthy of media attention, but in practice there is a wall of silence. The reason for this neglect may be that everyone is so familiar with the problem in themselves or in their friends and relatives that the unpleasant facts are ignored in favor of something new and the evanescent breakthroughs that enchant the press. It may also be an example of a taboo subject from which we cringe.

While silence reigns in public, some doctors have been paying close attention and some label lower back pain as an epidemic. Attacks of low back pain are usually of sudden onset. In a ten-year survey of all workers in the Boeing aircraft factories, attacks were found to be equally common in shopfloor workers engaged in heavy lifting and in clerical workers whose occupation involved only light work. Some 80 percent of the victims had a relief of pain within two weeks but 10 percent were still in pain five months later. Even the brief episodes tend to recur and may become more frequent and prolonged. Very careful testing of people with sudden-onset lower back pain reveals up to 15 percent may have one of five disorders that may explain the pain: slipped discs, displaced vertebrae, overgrowth of bone, unstable vertebrae, and fractures, tumors, and infections. This leaves 85 percent of the victims in the highly unsatisfactory category of "nonspecific lower back pain."

Many countries have set up commissions to give official guidelines on how to cope with these people. The most recent and infamous is entitled *Back Pain in the Workplace: The Management of Disability in Nonspecific Conditions* and was written by an international team of experts. Society intrudes to form an unholy coalition of employers, insurance companies, lawyers, and workers' compensation departments with puzzled doctors. They emphasize the ruinously rising costs of lost work hours, health benefits, and insurance plans. Because the doctors could define no traditional cause for the pain and disability, many of the inspecting alliance were eager to turn to the attitude of the victims as the cause of their pain. A recent survey of the British civil service showed absenteeism to be relatively low among the top-grade executives and the lowest grades, such as postmen, whereas the middle ranks, who face daily hassles, had the highest rate. The Boeing survey identified job dissatisfaction as predicting those most likely to complain to the company of back pain. It seems to me such obvious common sense that those who hate their job and the company will complain to the company. It is hardly worth the trouble of a vast survey.

Ignoring this commonsense explanation, the report on back pain in the workplace concludes that dissatisfied workers cause their own pain. In order to treat this common problem, the report proposes specific treatment. For the first six weeks, the victim of nonspecific lower back pain is permitted only a day or two of bed rest, after which movement is vigorously encouraged with professional help and with minimal analgesic medication. It is quite true that the majority recover during this conservative regime, at least until their next episode. The commission is even more specific about the proper treatment of those still in pain after six weeks. The diagnosis of back pain is to cease and the patients are now to be labeled "movement intolerant." I take this phrase to be a politically correct neologism implying a work-shy shirker. Furthermore, it recommends a cessation of all further medical treatment on the grounds that it positively encourages patients to consider themselves sick. In order to reinforce this, it proposes an abrupt end to the payment of health benefits and the relabeling of workers as unemployed.

This report is the considered opinion of a very eminent international grouping of establishment experts. They conclude that the problem is no longer for traditional medicine but is instead a social, psychological epidemic and should be treated as such. The Canadian Pain Society objected strongly to the report. All societies contain large numbers of "experts" who have precisely diagnosed the causes of what is wrong with their societies. Immigrants, minorities, and criminals are popular explanations for society's ills.

One popular idea is that society is sinking under a mass of people who live a life of ease and luxury supported by social benefits. The Australian psychologist Issy Pilowsky invented the term *hypochondriophobia* to label the tendency in our population to suspect and fear the validity of people on prolonged disability benefits. For example, Italians love to repeat a press fabrication about a man who was on a pension for the blind while also being paid as a traffic policeman. In this atmosphere, in which social and disability benefits are considered mainly in terms of cheating, fraud, hypochondria, and lack of moral fiber, governments concentrate on ways to reduce their social security budgets. This is not a good atmosphere in which to mobilize the mass of good-hearted citizens who would love to take part in social action to help and encourage the lonely abandoned folk who live in pain.

Epilogue

This book has been about the many challenges of the many pains. Inevitably, I have kept returning to the urgent practical question of how to control pains. But beyond that question lie deeper ones, and the practical question will not be answered satisfactorily until we understand more of the context in which pain resides. Pain is one facet of the sensory world in which we live. It is inherently ridiculous to consider pain as an isolated entity, although many do exactly that. Our understanding brains steadily combine all the available information from the outside world and within our own bodies with our personal and genetic histories. The outcomes are decisions of the tactics and strategies that could be appropriate to respond to the situation. We used the word *pain* as shorthand for one of these groupings of relevant response tactics and strategies. Pain is not just a sensation but, like hunger and thirst, is an awareness of an action plan to be rid of it.

Index

CPSIA information can be obtained
at www.ICGtesting.com
Printed in the USA
LVOW13s1105230817
546069LV00005B/17/P

9 780231 120074